Management for Professionals

More information about this series at http://www.springer.com/series/10101

Oliver Busch
Editor

Programmatic Advertising

The Successful Transformation to Automated, Data-Driven Marketing in Real-Time

Editor
Oliver Busch
Hamburg
Germany

Translation from the German language edition:
"Realtime Advertising" by Oliver Busch
Copyright©Springer Gabler 2014
Springer Gabler is part of Springer Science+Business Media
All Rights Reserved

ISSN 2192-8096   ISSN 2192-810X   (electronic)
Management for Professionals
ISBN 978-3-319-25021-2   ISBN 978-3-319-25023-6   (eBook)
DOI 10.1007/978-3-319-25023-6

Library of Congress Control Number: 2015957245

Springer Cham Heidelberg New York Dordrecht London
© Springer International Publishing Switzerland 2016
This work is subject to copyright. All rights are reserved by the Publisher, whether the whole or part of the material is concerned, specifically the rights of translation, reprinting, reuse of illustrations, recitation, broadcasting, reproduction on microfilms or in any other physical way, and transmission or information storage and retrieval, electronic adaptation, computer software, or by similar or dissimilar methodology now known or hereafter developed.
The use of general descriptive names, registered names, trademarks, service marks, etc. in this publication does not imply, even in the absence of a specific statement, that such names are exempt from the relevant protective laws and regulations and therefore free for general use.
The publisher, the authors and the editors are safe to assume that the advice and information in this book are believed to be true and accurate at the date of publication. Neither the publisher nor the authors or the editors give a warranty, express or implied, with respect to the material contained herein or for any errors or omissions that may have been made.

Printed on acid-free paper

Springer International Publishing AG Switzerland is part of Springer Science+Business Media (www.springer.com)

Preface

Volatility [vollaltillilty] (variability, instability) nicely sums up the most radical change that has taken place in the global economy over the last decade and is certainly more apt than the word *digitalization*. As business people and consumers, we are seeing large companies and even entire markets rise and fall at ever shorter intervals, with developments that once took several decades now playing out within the space of 10 years. This increasing pace looks set to continue as physical goods manufacturers optimize their processes based on ever shorter innovation cycles. Digital product makers have already adopted agile working methods by moving away from major releases towards weekly and sometimes even daily launches. It took 10 years for the digital revolution to shift conventional business processes and models to the stationary web. Now we are again witnessing another rapid shift, but this time the web is moving away from desktop to mobile devices. Always on – everywhere – infotainment on the go with sight, sound and motion.

Flexibility and innovation are both essential to surviving in these evolving conditions in which the fastest-growing companies over the last decade have managed to turn volatility into success by offering their customers innovation upon innovation, by making sure every single thing they do is dynamic and by adapting time and time again to the permanent change going on around them. They adopt a flexible approach to opportunities, ideas and extrinsic changes that present themselves. Conventional organizations with long-standing structures and planning processes appear to be too sluggish when it comes to turning rapid technical progress into relevant customer benefits. As the adage goes, better remains the enemy of good. And the next better now lands on your doorstep faster than ever before.

Everything that applies to markets and businesses is, of course, also highly relevant to marketing as market-oriented management. Volatility is a huge challenge for marketers as they need to get ahead of the wave and drive the markets before they get driven, which in turn requires them to turn to innovation and flexibility. They have to use increasing levels of technology and digitalization to carve out a constant edge over the competition, while also being able to perform complex tasks before others learn how to do them.

Innovation should not be an end in itself, and taking action simply for the sake of it is just as uncalled for as concepts of yesteryear. There is always going to be something that is better to at least try out and do badly rather than not trying to do it at all. Great marketers need to develop a feel for cast-iron constants, in the same way as they need to be constantly open to decisive, disruptive developments. This applies to brand management, brand communications, campaign management and advertising impact monitoring. The evolutionary speed of artificial intelligence is helping to drive corporate information processing and decision-making, while also presenting new opportunities for better marketing each and every day. The human brain, which naturally includes the consumer brain, is not subject to this acceleration. This in turn means that the basic principles of consumer psychology and advertising impact research are not affected by volatility, thus also increasing demands on a perfect CMO.

Programmatic advertising is both the keyword and subject of this book, a term that brings together the developments outlined above as well as a catalogue of technical developments from the past decade. According to the McKinsey Global Institute, aside from the mobile web, the operationalization of 'big data', machine learning and the automation of knowledge-based decisions are the most significant disruptive changes expected to take place over the next decade. Marketing communication is currently just a prelude to a more comprehensive and renewed upheaval of all marketing disciplines that will give companies access to modern systems allowing them to make millions of specific decisions a day based on dynamic, real-time information. And this is just preparation for the next level, artificial intelligence-driven marketing. But first things first though – we should start with the principle of programmatic advertising.

This book gives marketers and media workers the tools they need to actively shape the shift that has already started, to embrace and implement innovation and flexibility and to make friends with volatility. Readers will learn the basics and circumstances surrounding strategy development while also being able to enter into informed conversations and negotiations with new and existing business partners. Innovation is rarely the result of an individual, and this is also the case with this book, as it is a global effort consisting of 45 well-respected experts on the subject from the USA, EMEA and APAC. All that remains is to make sure things remain dynamic, so feel free to follow, share and discuss ongoing news and views with the authors and other readers by visiting www.facebook.com/programmaticadv.

Hamburg, Germany   Oliver Busch
July, 2015

Contents

Part I Concept

The Programmatic Advertising Principle 3
Oliver Busch

Borderless Media Management 17
Rosa Markarian, Aee-Ni Park, and Mark Grether

Programmatic Disruption for Premium Publishers 25
Holm Münstermann and Peter Würtenberger

Perspectives of Programmatic Advertising 37
Jürgen Seitz and Steffen Zorn

Part II Components

Consumer-Centric Programmatic Advertising 55
Oliver Gertz and Deirdre McGlashan

Understanding Demand-Side-Platforms 75
Arno Schäfer and Oliver Weiss

Granularity Creates Added Value for Every Objective 87
Arndt Groth and Viktor Zawadzki

Enhanced Success with Programmatic Social Advertising 103
Patrick Dawson and Michael Lamb

Programmatic Brand Advertising 111
Stephan Noller and Fabien Magalon

The Creative Challenge 123
Sven Weisbrich and Caroline Owens

Unleashing the Power of Greater Creatives for Brands 131
Chip Meyers and Christian Muche

Cross-Channel Real-Time Response Analysis 141
Burkhardt Funk and Nadia Abou Nabout

The Contribution of Measurement in a Cross-Device, Data-Driven, Real-Time Marketing World 153
Niko Marcel Waesche, Tilman Rotberg, and Florian Renz

How to Be a Successful Publisher in the Programmatic World 165
Frank Bachér and Jay Stevens

Part III Transformation

The CMOs Challenge 179
Ralf E. Strauss and Jonathan Becher

Integrated Campaign Planning in a Programmatic World 193
Andy Stevens, Andreas Rau, and Matthew McIntyre

Evolution of Digital Campaign Design and Management 211
Nils Hachen and Stefan Bardega

Realtime Data Accelerates Online Marketing 221
Kolja Brosche and Arun Kumar

Redefining Retargeting 233
Grégory Gazagne and Alexander Gösswein

Driving Performance with Programmatic CRM 243
Florian Heinemann

Pricing for Publisher: Scaling Value, Not Volume 255
Marco Klimkeit and Paul Benson

Managing to Quality Attention and Outcome Through Programmatic Technology 265
Ted McConnell and Lothar Hoecker

For Social Good 279

List of Contributors

Frank Bachér The Rubicon Project, Hamburg, Germany

Stefan Bardega ZenithOptimedia, London, UK

Jonathan Becher SAP, Palo Alto, CA, USA

Paul Benson Adition UK, London, UK

Kolja Brosche TheAdex GmbH, Frankfurt, Germany

Oliver Busch Facebook Germany GmbH, Hamburg, Germany

Patrick Dawson MediaMath, London, UK

Burkhardt Funk Leuphana Universität Lüneburg, Lüneburg, Germany

Grégory Gazagne Criteo, Paris, France

Oliver Gertz MediaCom, Munich, Germany

Alexander Gösswein Criteo GmbH, Munich, Germany

Mark Grether Xaxis (A GroupM Company, Part of WPP), New York, NY, USA

Arndt Groth PubliGroupe Ltd, Lausanne, Switzerland

Nils Hachen Zenithmedia GmbH, Duesseldorf, Germany

Florian Heinemann Project A Ventures GmbH & Co. KG, Berlin, Germany

Lothar Hoecker Dentsu Aegis Network Germany, Wiesbaden, Germany

Marco Klimkeit Yieldlab AG, Hamburg, Germany

Arun Kumar Cadreon, New York, NY, USA

Michael Lamb MediaMath, New York, NY, USA

Fabien Magalon LiveRail, Paris, France

Rosa Markarian Xaxis (A GroupM Company, Part of WPP), Duesseldorf, Germany

Ted McConnell Ted McConnell Consulting Llc, Cincinnati, Ohio, USA

Deirdre McGlashan MediaCom, London, UK

Matthew McIntyre uniquedigital UK, London, UK

Chip Meyers Reactx, Santa Monica, CA, USA

Christian Muche dmexco, Cologne, Germany

Holm Münstermann Axel Springer Media Impact, Berlin, Germany

Nadia Abou Nabout WU Vienna University of Economics and Business, Vienna, Austria

Stephan Noller nugg.ad AG, Berlin, Germany

Caroline Owens IPG Mediabrands, London, UK

Aee-Ni Park Xaxis (A GroupM Company, Part of WPP), Duesseldorf, Germany

Andreas Rau uniquedigital GmbH, Hamburg, Germany

Florian Renz GfK, Hamburg, Germany

Tilman Rotberg GfK, Nuremberg, Germany

Arno Schäfer 161MEDIA B.V., Hamburg, Germany

Juergen Seitz Hochschule der Medien (HDM) Stuttgart, Stuttgart, Germany

Andy Stevens Syzygy, London, UK

Jay Stevens The Rubicon Project, London, UK

Ralf E. Strauss German Marketing Association, Hamburg, Germany

Niko Marcel Waesche GfK, Nuremberg, Germany

Sven Weisbrich UM – Universal McCann GmbH, Frankfurt, Germany

Oliver Weiss Adform Germany GmbH, Hamburg, Germany

Peter Würtenberger Axel Springer SE, Berlin, Germany

Viktor Zawadzki Spree7 GmbH, Berlin, Germany

Steffen Zorn Swinburne University of Technology, Karrinyup, WA, Australia

Part I
Concept

The Programmatic Advertising Principle

Oliver Busch

It is no secret that the digital marketing industry loves innovation and is thus in a constant state of change and always on the lookout for the next big thing. You could even go so far as to say this quest is often considered a business model itself. Nevertheless, this topic has another completely different and more extensive dimension to it, one that has been filling congress halls and schedules time and time again for years: data-driven marketing automation. If you take the time to look into the principles and aims of programmatic advertising (also known as real-time advertising), you will soon see why this is set to continue captivating the cross-media creative and media industry over the coming years, although this is increasingly likely to happen during daily business rather than at presentations and talks given at congresses. In the end we will see a transformation that permeates every discipline and business, and a scenario where programmatic advertising forms the basis of advertising and marketing on every level.

1 Background

There are a lot of terms surrounding programmatic advertising. Buzzwords such as big data and data-driven display refer to more sophisticated data used as a basis for making marketing and media decisions, while other keywords such as machine learning or prospecting focus more on algorithmic data processing and the statistical processing of continuously generated empirical data. Programmatic buying and automated trading constitute another area of focus within this context, while the narrowly defined term real-time bidding (RTB) is mainly prevalent in Anglo-American markets and stresses the option to determine prices dynamically through auctions. What this all boils down to is a more intensive use of data, technology and

O. Busch (✉)
Facebook Germany GmbH, Caffamacherreihe 7, 20355 Hamburg, Germany
e-mail: ollibusch@fb.com; ollibusch@me.com

© Springer International Publishing Switzerland 2016
O. Busch (ed.), *Programmatic Advertising*, Management for Professionals,
DOI 10.1007/978-3-319-25023-6_1

artificial intelligence in marketing with the common goal of boosting marketing efficiency – in real time.

1.1 Requirements

Solving a particular problem is a key characteristic that sets a long-term marketing trend apart from the multitude of hyped and short-lived innovations. Given the attention paid to programmatic advertising, the level of suffering in the digital advertising industry has to be just as high as the expectations people have on it improving the status quo. And it would seem that both cases apply here.

Associations of National Advertisers around the world think that the digital industry has reached a dead end where there is no transparency, a lack of measurability, unviewable ad impressions and even fraud. As a result of these factors and a lack of determination to go digital, advertisers are putting their customers off from following them into the digital world (Sluis 2015). However, customers do not just want to see proof of performance. Online advertising is also subject to the basic economic principle whereby constant efficiency gains are required in order to boost productivity and competitiveness. Major multinational advertisers in particular are facing immense pressure to optimize their processes due to globalization. If no tangible performance improvements in brand advertising are achieved, prices are simply dropped even more resolutely. In doing so, list prices remained stable over the years, but the gap between gross and net earnings widened as time went by. Downscaling online editorial teams due to insufficient capitalization now appears to have reached its natural boundaries, which goes to show that further price reductions do not represent a solution to the problem of brand building performance.

A similar scenario can also be seen in performance marketing, i.e. among sales-oriented online advertisers. Competition in the ecommerce domain is fierce with everyone battling it out to become the (niche) market leader since there is no such thing as a solid second place in the online world. Marketing efficiency and the battle for top rankings go hand in hand with carving out market share gains in a highly competitive environment defined by minimal margins. An audience that returns good results is defended to the brink of profitability. An adequate assessment of effects throughout the customer journey is absolutely essential to avoid making false estimates and incorrectly allocating resources. However, investors do not want to see gradual optimization in the online business, they want to see strong growth. A surplus of advertising space on the stationary web and mobile devices enabled poor efficiency to be balanced out by low prices and high volumes. But this is no way to achieve genuine scalability, especially as the technical costs of serving ads and measuring success often exceed the value of low-cost ad space.

The media are all too familiar with the pressure to grow. Traditional media's financial reserves continue to dwindle, which naturally puts conventional marketing firms under increasing pressure to capitalize their online segment. Only a few media have managed to "premiumize" their advertising space offerings in which the perceived quality of the advertising environment is also reflected in a price premium in the increasingly user-centric ad tech world (Loechner 2015). But even when faced with this increasing pressure to perform, premium status is merely a temporary victory that is unlikely to succeed in the long term if not developed further in the future. Price pressure is aggravated by the amount of work involved in booking and processing digital advertisements. Unlike TV and print, the online advertising industry is years behind its former analog counterparts when it comes to process automation. To put it pessimistically, almost 20 years after the onset of the digital revolution, the entire process from planning to booking and optimization is still dominated by manual processes that involve Excel sheets, faxes and post-its.

The urgent need to boost efficiency brings purchasing and sales together, and it is this united effort that is helping to drive the currently observed quest for new technology.

Whenever talking about the urgent need for change, it is important to bear the main stakeholders, i.e. consumers, in mind when doing so as their use of media is currently changing faster than the marketing industry. More and more people are using their mobile devices to consume media (eMarketer 2014a), even in areas not typical to mobile, which is pushing traditional media to the back of the cupboard – even at home (eMarketer 2014c). In contrast to desktop use of the stationary web, user experience on smartphones and tablets focuses on discovery rather than search. People start out by browsing their social media apps where billions of users go online to read, post, like and comment on updates of their friends, preferred media, favorite stars and brands all the way from breakfast time through to turning off the lights at night. Users swipe away on their device's screen with their thumb to browse the plethora of content while their brain sifts out the nuggets of information that are of relevance to them. Anyone who used to think the remote control was a threat to commercial breaks probably has heart palpitations then thinking about the current situation for advertisers. Core media has never been closer to consumers, as personal as it now is with native ads embedded in content. Nowadays recipients vehemently defend themselves against poor, irrelevant advertising – if they are unable to block it out of their mind, they will take more radical action and use an ad blocker. If a user's thumb is supposed to pause for an advertising post in between the headlines of the New York Times and an update posted by a good friend, the publisher and the brand need to do everything they can to ensure individual relevance across all of the devices in use. As a result of this situation, media use is shifting towards offers that are extremely relevant at a specific moment in time. Messages that follow this principle will grab people's attention, and it seems that recipients themselves want to see a boost in efficiency as people's time and attention spans are limited.

This unifying need explains the urgency with which the entire ecosystem is driving the change towards programmatic advertising.

1.2 Development

The flood of information on the marketing scene may cause some people to think that programmatic advertising is just a hype that popped up out of nowhere. However, programmatic advertising actually emerged from a number of different developments initiated well over 5 years ago and only became a game-changer when these developments came together:

- Immense computing power:
  Complex data needs to be computed in milliseconds in order to allow for real-time marketing decisions.
- Inexpensive data storage:
  Millions of anonymized stored data points can potentially be used to forecast campaign-specific values for each individual ad space.
- Science in marketing:
  New technology allows marketers to persuade mathematicians and quantum physicists to join them in their work.
- Stock exchange structures when trading ad space:
  Pricing based on real-time supply and demand for each individual ad impression opportunity at the moment it is created helps to make markets more dynamic.
- Globalization of advertising markets:
  Advances in standardization, a global infrastructure and globally operating market players help to speed up the pace of development.
- Fast data connections
  Correspondence on how to get the most out of an individual ad impression can be sent around the world several times over within a matter of milliseconds.
- Personalization
  "Real value comes with real identity" is the basic principle now banishing the pure anonymity seen on the web during its infancy. Log-ins and permissions enable dialogs to be conducted across multiple devices.

These developments all come together to provide new opportunities that the marketing industry is currently exploring in order to solve the problems mentioned previously. Huge sums of money are being made available due to the advertising market's interest in this science and technology, and this is set to drive and speed up the process as a whole. Additional rounds of financing give platform providers access to hundreds of millions of euros in expansion capital (Kroll 2013), which in turn boosts the rating of the service providers and media involved. And rightly so, it would seem, given current growth forecasts and the path chosen by leading global advertisers: "Technology is providing a huge boost to marketing productivity, so we need to understand marketing automation." Tom Buday, CMO at Nestlé, said when describing the task (TheDrum 2014b). "We see programmatic as an opportunity to take a massive investment that we have, which is media, and pull data from that investment. So it is no longer just a dumb investment." explained Bonin Bough, VP Global Media at Mondelēz (TheDrum 2014a), who at the same time is breaking up the media silos. Marc Pritchard, Global CMO at P&G, is also resolute about this transformation: "The thing about programmatic, it's just inevitable, because of the power of technology ... and at the heart of it to make it a better experience for consumers." (Neff 2015) Media expert Sascha Jansen from Annalect aptly summed up the situation as follows: "Maybe not in 5 years, but perhaps in ten we will look

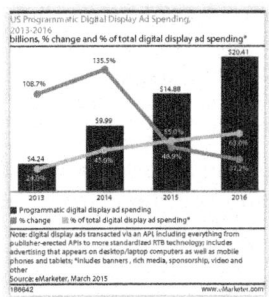 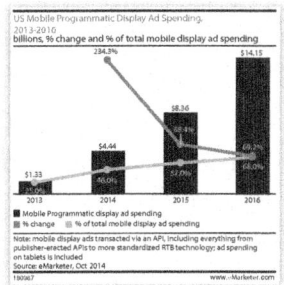 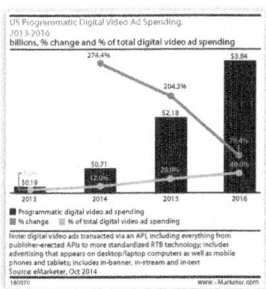

Charts 1–3 US programmatic ad spending on display and video as well as mobile (eMarketer 2014a)

back and laugh at how we used to plan online media." (Zunke 2013). Having said that, we may well see all this happen much sooner (Charts 1–3).

2 Classification

An analysis of background and origin is crucial to strategically locating these new opportunities. As always, this depends on the viewpoint and position within the market. Advertisers, agencies, the media and sales houses consider programmatic advertising to be both an opportunity and a threat as it appears to be highly complex, yet it also appears to reduce complexity when striving to achieve ambitious business development goals. Programmatic advertising presents the marketing industry with new opportunities, but should not be mistaken as being an isolated discipline or instrument to be observed separately. Classifying programmatic advertising as a new marketing or advertising sales channel, as a new set of tools, or even as a new pricing method for trading digital advertising space would be tantamount to voluntary self-restraint because the term actually describes a new value-adding principle that can be applied to digital advertising against the backdrop of the insuppressible digitalization of advertising and marketing. Nevertheless, it remains a matter of choice as to whether or not this can be considered an evolution or a revolution.

2.1 Definition

We always seem to have difficulties in defining developments in their early stages, as was the case when classifying online marketing at the start of the millennium (Busch 2000). Terms such as "excessive" and "unrealistic" were often used when experiencing initial difficulties, yet just a few years later it turned out that people were thinking on way too small a scale. With this in mind, programmatic advertising should have as open a definition as possible in terms of its core characteristics, while also providing sufficient scope to allow affected marketing and marketing communication disciplines to get involved and experiment within their field of work.

Below are the main characteristics of the programmatic advertising principle:

1. Granularity,
   i.e. full consideration of individual ad impression opportunities together with their general parameters, specific recipients and specific advertising environment.
2. Real-time trading,
   i.e. deciding on a specific advertiser or a specific ad impression opportunity at the time of its creation and based on the latest data.
3. Real-time information,
   i.e. assessing the available opportunity based on its highly specific characteristics and relevant empirical data collected to date.
4. Real-time creation,
   i.e. advertisers serve a (where possible: dynamic and data-driven) ad that is best suited to the opportunity immediately after winning the bid.
5. Automation,
   i.e. an automated booking and posting process.

In a nutshell:

▶ "Programmatic advertising describes the automated serving of digital ads in real time based on individual ad impression opportunities" (Bardowicks and Busch 2013).

In view of the problem set out above, the programmatic advertising principle can be boiled down to *granularity* and *automation*. Granularity provides advertisers with a new way to optimize budget efficiency as individual ad impression opportunities can be selected, evaluated, priced and created at an unprecedented level of specificity thanks to comprehensive scientific forecasting methods. Automation opens the door to this highly specific way of looking at and dealing with individual ad impression opportunities as it permits a far more extensive decision-making process for each individual ad impression than humans were able to achieve to date during the course of entire campaigns. Unlike the previous train of thought, the aim here is to create long-term value rather than focus on low prices in the form of real-time bidding as advertisers are only prepared to pay higher prices if they see greater value, which of course benefits the profitability optimization aims of everyone involved. The automation of complex manual processes also promises cost savings, although this cannot be expected for the time being as a result of having to refinance the necessary systems. The additional charges imposed by system providers already have to be compensated by the added value they generate in order to justify their use. A few globally scaled media technology companies represent an exception to this as they provide data as well as proprietary end-to-end booking interfaces without any additional charges, thus making it easier for advertisers to get started.

2.2 Functionality

Facebook and Google were both early adopters and promoters of systems that provide millions of advertising customers with the opportunity to manage their automated, data-driven campaigns in real time, primarily on Facebook and Google pages. Third-party programmatic advertising technology providers have adopted the main basic principles from here, namely the auction mechanism, granularity and transparency (cf. Heimann 2013). This not only bundles demand, it primarily concentrates the highly fragmented global offering of online media. This aggregation drastically reduces the number of players to be connected on the real-time advertising trading floor, which is what actually makes automated trading technically feasible in the first place. For this to happen, the market players need global standards for protocols, ad formats and methods so that they can tap into the global growth potential.

Media and their sales houses – who provide the advertising market with their ad impressions in real time – use their own interfaces (API), as do Facebook and Google, or hook up to a supply-side platform (SSP), also known as a sell-side platform. It provides media with all (or selected) ad impression opportunities as well as accompanying details of each (or selected) advertisers based on the specified conditions. During the early adoption phase, advertising space marketers generally only used SSPs to auction off (cf. RTB) low-demand, left-over space. But nowadays SSPs are increasingly reproducing the complex marketing models used in conventional sales processes, but subsequently refining them by incorporating analyses. Private deals, also known as automated guaranteed, are now also a key component and include prior negotiation of the quantity, terms and sometimes also priority to look into it. Otherwise, private deals are handled according to the programmatic advertising principle by sending a deal ID via the requisite systems.

Ad impression opportunities made available this way are received at their time of creation by so-called demand-side platforms (DSPs) and then evaluated according to predefined selection criteria, empirical data and forecasts for each of the designated advertisers. Cookie-based DSPs are accompanied by an analog world of providers who manage multi-device advertising based on log-ins to corresponding (social) media. These are collectively known as Facebook Marketing Partners (FMPs) based on their main, but not only, inventory source.

Advertisers and their agencies stipulate their terms just as sales houses do, and then an automated decision is taken as to whether value and price can be brought into line with one another. If an ad impression is traded via auction, the advertiser who attributes the highest value to a specific opportunity and thus bids the highest value will win the auction. If the price and right of first opportunity were agreed with the advertiser in advance, the DSP/FMP will calculate in real time whether or not the bids match the impression opportunity and whether the agreed price is worth it. The entire process from creating an ad impression opportunity to evaluating, selling, processing and serving an ad to users generally takes approximately 50 ms,

Fig. 1 Schematic illustration of an advertising space auction via SSP and DSP in approximately 50 ms (Zawadzki 2013)

and this process (cf. Fig. 1) generally happens millions of times a second via DSP and SSP or FMP and API.

2.3 Components

Anyone interested in an comprehensive and organized overview of every programmatic advertising component provider will probably refer to the "Display LUMAscape" and "Social LUMAscape" charts provided by technology consultancy Luma Partners LLC (Kawaja 2014). The downside to these well-known charts is their inevitable level of complexity. Having said that, if you were to map out every conceivable ingredient of bread along with every supplier present on the market, the resulting chart would not be any less complex. The vast number of alternatives are not important here, but rather the basic interaction between the main components, which can be seen below in simplified form (Fig. 2).

The main players in programmatic advertising are the primary trading partners, i.e. advertisers and advertising media, both of whom are still represented by their

The Programmatic Advertising Principle

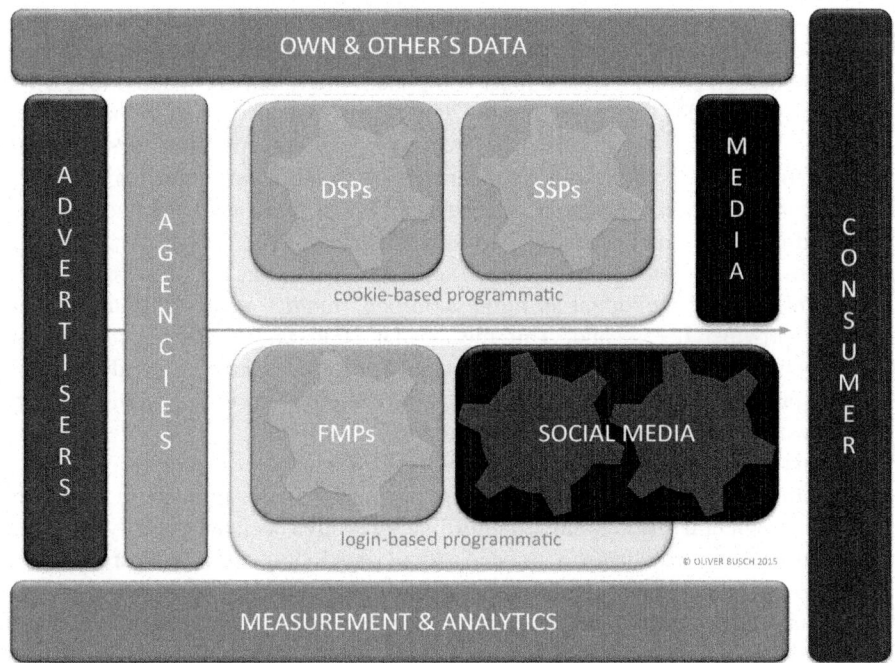

Fig. 2 Primary components of programmatic advertising (own image)

purchasing and sales specialists, namely agencies and sales houses. In spite of the inevitable debates at the outset, programmatic advertising does not yet appear to have triggered a reduction in the value chain as a result of insourcing sales and procurement tasks. In fact these new options are generating more appreciation for both disciplines due to the increased need for certain know-how, experience and guidance, which in turn gives rise to special service providers who can temporarily step in to fill the gap.

SSPs and DSPs are two of the three core components needed to get access to aggregated supply as well as aggregated demand. The main providers of both are connected to all of their counterpart's providers via interfaces, each numbering around a dozen. The resulting virtual marketplace can then in theory send any offer to any client. Facebook also has one of these cookie-based SSPs in the form of Facebook Exchange, but log-in-based programmatic ad impressions are generally traded directly between the FMP platforms and the interfaces of the respective social media offering. Projects to bridge the gap between the two programming worlds are already in full swing: on the hand DSPs, FMPs and search-bidding providers are being acquired and integrated, while, on the other hand, Facebook provides anonymized usage data via its ATLAS Ad Server, which advertisers can log into from a range of different devices to manage and measure digital advertising both on and outside of the Facebook platform.

These core components are now also flanked by a broad ecosystem of supporting components. In contrast to the major consolidation tendencies among core component providers, the number of support component providers is expected to grow in the long term (cf. Reisch and Rinderle 2013). This extended programmatic advertising ecosystem includes, in particular, data-management platforms (DSPs), data and targeting offerings, monitoring and validation tools, and advanced tracking and analysis features. Advertisers, media or both can take advantage of such components to help them optimize their automated tasks. This open DSP and SSP interface concept makes it easy to hook up optimization tools so that players can simply test any new opportunities using individual ad impression opportunities without having to liaise with the provider beforehand. This means that DSPs and FMPs are turning into multifunctional platforms where features that used to be booked and used separately are now provided under a single umbrella. Think of it as a recording studio's mixing desk to which new features can be added at regular intervals.

Given the huge number of ways to statistically evaluate measureable attributes in programmatic advertising, conventional performance values relevant to marketing such as incremental reach, brand lift or offline sales have been temporarily upstaged. This could be owing to the need to spend a lot of time looking into and learning about all the programmatic options, but it is more likely due to the more technical background of programmatic advertising pioneers who are far more familiar with an attribution model for a customer journey analysis than a conventional lift test involving a test and control group used to determine ad impact under the same external conditions. Reintegration of proof of ad impact on traditional research within a programmatic context will be dealt with in more detail later on in this book.

2.4 Scope

Over the last 20 years, no more than four developments had sufficient potential and power to grasp and influence marketing communication across every tool and discipline:

- Internet, which acts as a digital counterpart to the analog world,
- Social web, which is a digital map of links and relations between humans and objects,
- Mobile web, which serves users multimedia content and ensures they are "always on",
- Programmatic advertising, which generates additional value within digital marketing communication.

As with the other meta-trends previously mentioned, differing adoption speeds do not prevent the programmatic advertising principle from being limited to certain campaign goals or disciplines. Programmatic advertising is neither suited to branding goals, nor to performance – be it web or mobile, display or video advertising, not even media or creation. Positioning programmatic advertising as one of several sales channels for publishers or as part of a media plan is a sure-fire way of allowing the competition to maximize value-added potential. Instead, programmatic advertising should be seen as a principle that allows integrated, data-based communication within a fragmented media world, while also helping to clear inefficient marketing silos out of the way. It connects media with creation in order to serve the right people with the right messages at the right time, and all of that on a large scale. It links marketing disciplines and advances people through the marketing funnel. On top of that, programmatic advertising for media means having to develop offerings tailored to the needs of tomorrow's advertising customers, yet it also serves as a basis for refining inventories and implementing more sophisticated sales strategies.

While this exact implementation is currently taking place, programmatic advertising is already edging beyond the boundaries of the stationary and mobile web as its very nature prompts it to do so. Among other things, it is possible to tap into direct potential for programmatic advertising by linking offline and online data in a privacy-compliant manner with leading dialog marketing, micromarketing and household advertising providers. The digitalization of traditional media channels is also paving the way for expanding the principle to outdoor media, radio and TV advertising, and – to a limited extent – even to print advertising (Schulze-Geissler 2013; Roloff 2013). While Programmatic TV is still a vision, it is by no means a fairytale: video advertising is currently conquering internet-capable devices, while the boundaries between online and offline video are disappearing rapidly thanks to non-linear offerings from Netflix & co., mobile content broadcasted to stationary devices (AirPlay), and smart TVs that can connect to the internet. There will be almost 80 million smart TV users in the US by 2018 (eMarketer 2014b), while in some European countries twenty percent of TV owners already have a smart TV (eMarketer 2015). This represents a major change to the video advertising environment as social media networks, video-on-demand platforms, streaming providers and smart TV manufacturers compete with TV stations and cable network operators by offering advertisers a wide range of programmatic benefits. "TV inflation is up, you get great reach and coverage, but if I can run the ads programmatically then I get display and video efficiency too," stated Gawain Owen, Nestlé UK, "I will pay more for programmatic, I might even pay two times as much because it drives efficiencies," (McEleny 2015).

According to ecommerce pioneers, the main development drivers are now global advertisers and agencies. Ex-IPG Mediabrands CEO Matt Seiler recognized this and came up with the ambitious goal of automating all US media buying – not just digital – to 50 % by 2015 (O'Leary 2014). "This really is a fundamental change," explained Marc Panic, COO at GroupM Interaction EMEA. "It is going to extend across every digital channel and further afield as offline media are going digital and the standard course taken in media trading is search, display, video and now mobile. TV, radio and outdoor are next on the list" (Knapp and Marouli 2013).

Marketing disciplines that have drifted apart are highly likely to converge again due to data and automation. This not only applies to online and offline advertising or conventional and dialog marketing, but also to media and creation as dynamic, flexible and real-time advertising becomes an integral part of the programmatic advertising principle and many of the corresponding systems. The data sources used to select the advertising space may also serve as the basis for selecting or creating the right media for the ad impression opportunity. Art directors and copywriters are still unaccustomed to coming up with dynamic creative concepts in line with brands, but this is starting to change thanks to creative media ideas that react in real time to data from their surroundings. Ultimately this will lead to stronger ties between these contrasting professions.

3 Conclusion

No-one in procurement or sales will talk about programmatic advertising in 5–10 years' time as the programmatic principle will form the basis of all marketing communication – irrespective of the media – as automation and data will be closely linked to other marketing disciplines.

During this transformation, the individual market players' programmatic advertising adoption and learning speed will lead to major market share shifts, both in terms of media and on the part of advertisers. From now on, technology will no longer be a minor point of discussion and taken to be a generally available basis for more conceptually and creatively demanding integrated marketing communication.

Bibliography

Bardowicks, B., & Busch, O. (2013). Diskussionspapier: Programmatic Advertising. bvdw.org, http://www.bvdw.org/medien/bvdw-diskussionspapier-beleuchtet-entwicklungen-im-realtime-advertising?media=5002. Published on 12 Aug 2013, p. 4.

Busch, O. (2000). Markenführung im Digital Age, in: absatzwirtschaft marken. Published on 1 June 2000, p. 24 et seq.

eMarketer. (2014a). Mobile continues to steal share of US adults' daily time spent with media. eMarketer, http://www.emarketer.com/Article/Mobile-Continues-Steal-Share-of-US-Adults-Daily-Time-Spent-with-Media/1010782. Published on 22 Apr 2014.

eMarketer. (2014b). Majority of US internet users to use a connected TV by 2015. eMarketer, http://www.emarketer.com/Article/Majority-of-US-Internet-Users-Use-Connected-TV-by-2015/1010908. Published on 13 June 2014.

eMarketer. (2014c). Second screening during TV time—It's not what you think. eMarketer, http://www.emarketer.com/Article/Second-Screening-During-TV-TimeIts-Not-What-You-Think/1011256. Published on 6 Oct 2014.

eMarketer. (2015). Digital device penetration in the US and select countries in Western Europe, Nov 2014. eMarketer, http://totalaccess.emarketer.com/Chart.aspx. Published on 5 Mar 2015.

Heimann, T. (2013). Supply-Side-Plattformen – ein Leitfaden für die Verkaufsseite. In BVDW (Hrsg.) *Kompass programmatic advertising* (p. 16). Düsseldorf: BVDW. Published in September 2014.

Kawaja, T. (2014) Display LumaScape, http://www.lumapartners.com/resource-center/lumascapes-2/. Publication date undisclosed.

Knapp, D., & Marouli, E. (2013). Adex benchmark 2012 – European online advertising expenditure. IAB Europe, Published on 28 Aug 2013, p. 19.

Kroll, S. (2013). Mit neuem Kapital schneller wachsen. in: Internet World Business, http://www.internetworld.de/onlinemarketing/displaymarketing/neuem-kapital-schneller-wachsen-285523.html. Published on 25 Jan 2013.

Loechner, T. (2015). Premium publishers including Guardian, Reuters, FT launch programmatic alliance. MediaPost, http://www.mediapost.com/publications/article/245912/premium-publishers-including-guardian-reuters-ft.html. Published on 18 Mar 2015.

McEleny, Ch. (2015). Nestle's Gawain Owen: I would pay more for programmatic TV ads. *Marketing Magazine*, http://www.marketingmagazine.co.uk/article/1344197/nestles-gawain-owen-i-pay-programmatic-tv-ads. Published on 23 Apr 2015.

Neff, J. (2015). P&G's Pritchard on where marketing, media and metrics are going. Ad age, http://adage.com/article/cmo-strategy/p-g-s-pritchard-marketing-metrics/297592. Published on 16 Mar 2015.

O'Leary, N. (2014). Matt Seiler is out to remake and automate the media agency world. http://www.adweek.com/news/advertising-branding/matt-seiler-out-remake-and-automate-media-agency-world-155047. Published on 19 Jan 2014.

Reidel, M. (2013). OWM-Forderung: Ungewöhnlich scharfe Worte zum Dmexco-Start. HORINZONT.NET, http://www.horizont.net/aktuell/specials/pages/protected/OWM-Forderung-Ungewoehnlich-scharfe-Worte-zum-Dmexco-Start_116776.html. Published on 17 Sept 2013.

Reisch, S., & Rinderle, S. (2013). Real time bidding – Next level performance marketing. http://www.bluesummit.de/wp-content/uploads/2013/02/blueSummit-Whitepaper_Real-Time-Bidding.pdf. Publication date undisclosed, p. 13.

Roloff, F. (2013). Mobile programmatic advertising – ein Trend. In BVDW (Hrsg.) *Kompass programmatic advertising* (p. 41). Düsseldorf: BVDW. Published in September 2014.

Schulze-Geissler, A. (2013). Real-Time Advertising wird Außenwerbung beeinflussen. *Adzine*, http://www.adzine.de/de/site/artikel/9351/adtrading-rtb/2013/10/real-time-advertising-wird-aussenwerbung-beeinflussen. Erscheinungsdatum 23 Oct 2013.

Sluis, S. (2015). Procter & Gamble CMO Pritchard: Programmatic delivers business lift. Ad exchanger, http://adexchanger.com/brand-aware/procter-gamble-cmo-pritchard-programmatic-delivers-business-lift. Published on 6 Mar 2015.

TheDrum (2014a). Mondelez VP of global media, Bonin Bough talks internet of things, programmatic and tech. TheDrum, http://www.thedrum.com/file/nestle-cmo-tom-buday-talks-about-importance-technology-advertisers. Published on 18 Sept 2014.

TheDrum (2014b). Nestle CMO Tom Buday talks about the importance of technology to advertisers. TheDrum, http://www.thedrum.com/file/nestle-cmo-tom-buday-talks-about-importance-technology-advertisers. Published on 24 Sept 2014.

Zawadzki, V. (2013). A world in 50 ms. www.spree7.com/50ms. Publication date undisclosed.

Zunke, K. (2013). Die DSP gehört auf die "Demandside". ADZINE, http://www.adzine.de/de/site/artikel/7650/adtrading-rtb/2012/09/die-dsp-gehoert-auf-die-demandside. Published on 27 Sept 2012.

Oliver Busch joined Facebook in Hamburg in November 2013 as Head of Agency D-A-CH. His team works in collaboration with agencies to provide the foundation for sustainable success for businesses on Facebook. Oliver has been active in cross-media marketing for 15 years. In the past, he held various leadership positions in agencies as well as for marketers and advertisers. Furthermore, he has been engaged in various industry initiatives and organizations over the last decade and is co-founder and vice-president of the German iab's (BVDW) programmatic advertising group. Digital transformation and innovation adaption in the highly quality-conscious German-speaking markets are his central themes as an author, speaker and moderator.

Borderless Media Management

Rosa Markarian, Aee-Ni Park, and Mark Grether

Media management traditionally takes place locally in individual markets. Revolutionising the digital online market with programmatic advertising platforms creates an entirely new approach to managing online campaigns without being tied to a specific location. This article provides insight into how inventory and data management are changing as a result of "borderless" media management.

1 The Significance of Borderless Digital Advertising Media Commerce

As a result of globalisation and e-commerce, the whole world has turned into a huge department store for each and every one of us. Online shops all have the opportunity to offer their products to clients all over the world. Access to products is guaranteed across the globe. For example, if a German consumer looks worldwide for a particular product and finds it in the USA, the product can be ordered and delivered to that consumer's respective country. Nowadays, the capacity for this type of borderless commerce is taken for granted by consumers and business of all types worldwide. In the specific field of media management we now see an international trade of media inventory and of data. These two components play a key role in programmatic advertising and make borderless media management quick and effortless.

R. Markarian (✉) • A.-N. Park
Xaxis (A GroupM Company, Part of WPP), Derendorfer Allee 10, 40476 Duesseldorf, Germany
e-mail: rosa.markarian@xaxis.com; aee-ni.park@xaxis.com

M. Grether
Xaxis (A GroupM Company, Part of WPP), 132 W. 31st Street, 11th Floor, New York, NY 10001, USA
e-mail: mark.grether@xaxis.com

"Borderless media management means the offer and purchase of advertising space in countries other than those where the buyer or seller are based."

Automated media management has reached the global online market and is growing. According to eMarketer estimates, US programmatic digital display ad spend will grow 137.1 % to eclipse $10 billion in 2014, or 45.0 % of the digital display ad market. Significant growth will come from programmatic direct, expected to reach $8.57 billion in spending by 2016, or 42.0 % of the US programmatic market—up from 8.0 % in 2014. In Western Europe, research by IDC predicts that the Programmatic market will grow from $381 million in 2012, to $3.3 billion in 2017 and that spending on data technology and services in the region will hit $6.8 billion by 2018.

With programmatic advertising, an agency in any country can acquire inventory internationally – sourcing from the USA, Australia or Ghana is just as easy as it is to acquire inventory domestically. The following examples demonstrate the relevance of cross-border media management using programmatic advertising:

- Between January and March 2014, 65 % of buyers acquired inventory from more than one country using AppNexus – one of the world's leading programmatic advertising platforms. 42 % of all buyers even obtain inventory from more than ten countries on a daily basis.
- 98 % of inventory suppliers on AppNexus sell to buyers from different countries, with 87 % of inventory suppliers selling to buyers from at least ten different countries (AppNexus 2014).
- Xaxis – GroupM's Audience & Data supplier – operates in 19 countries across Europe. The bulk of its business is conducted locally in each country. In addition, a central team oversees international marketing teams for global clients and manages campaigns across different countries. Just five of the 19 countries generate more sales than the central team (Xaxis 2013).

Borderless media management already forms an integral part of programmatic advertising and is used in a number of markets.

2 How Digital Advertising Is Working Globally

Programmatic advertising is the result of a combination of technology and globalisation, with data and inventory playing key roles. The following section demonstrates the impact of globalisation in both areas.

2.1 The Inventory Aspect

The advantage of programmatic advertising is that inventory can be accessed quickly and easily with considerable reach. This gives buyers a wide selection and allows for very efficient processing. On the other hand, this also means that competition is very fierce for suppliers, given that marketers can offer inventory without any effort. These low entry requirements plus the initial use of programmatic advertising platforms largely for performance-oriented campaigns, may have contributed to the fact that, even now, programmatic advertising in the market is frequently equated with a market place for undifferentiated remnant inventory. However, as described in the opening section, programmatic advertising is a new, value-adding principle that covers all aspects of digital advertising. This is why those suppliers who can offer additional added value will be successful, both locally and globally. Indeed, this added value includes aspects such as high-quality inventory and advertising space, which are suitable for both performance and branding campaigns. Some other factors that move suppliers to the forefront are multi-channel products and implementation based on target group-specific data (targeting).

Brand safety tools are available in order to gain a better understanding of the quality of the large amount of inventory on programmatic advertising platforms. These tools analyse URLs and the content of websites and only display advertising on defined content. For technical reasons, however, the use of brand safety tools is no guarantee of 100 % safe implementation.

The key to offering automated premium inventory lies in so-called private marketplaces. In this model, specialist suppliers combine precise premium inventory and connect marketers directly to the programmatic advertising platform. Brand safety tools can then also be used on the pre-selected inventory. This combination of pre-selecting the inventory and automating content analysis offers quality advantages that are particularly attractive to major brand advertisers.

Another difference of private marketplaces is the fact that inventory is usually offered with a defined fixed price rather than using Real-Time-Bidding (RTB). Private marketplace suppliers buy and plan inventory in large quantities. This allows marketers to plan reliably and prices can be agreed that are below the RTB prices in the open market.

With private marketplaces, programmatic advertising can also be further developed to cover all areas of digital advertising. Inventory and advertising space considered premium or superior to standard ad spaces are often offered in private marketplaces before they are offered in the programmatic advertising open marketplace. Video demonstrates how far this development has gone. According to eMarketer, 34 % of marketers in the USA offer their premium video inventory directly on programmatic platforms. Video campaigns are already being implemented on a far-reaching scale using private marketplaces in Europe. In various markets, private marketplaces offer video services that cover 70–90 % of users in each country (Xaxis). Furthermore, mobile services on programmatic advertising platforms are still very sparse and will certainly be the next area to undergo significant further development through programmatic advertising.

Thanks to programmatic advertising platforms, it would be, from a technical point of view, very easy for any agency anywhere to implement a campaign in Australia. However, in practice insufficient expertise in local inventory quality and market conditions, combined with a lack of links with regional marketers, could lead to quality problems.

The advantage is therefore held by globally operating agencies that can establish private marketplaces with international marketers across different countries.

In comparison with local private marketplaces, transnational private marketplaces see advertising space being sold in larger volumes, which means that more price advantages are accessible to large global agencies than small-medium sized local agencies. Another advantage of global agencies is the ability to optimise international campaigns. In small to medium country specific agencies, optimisation expertise can be found locally, for example which advertising formats and approaches work best for which campaigns but these are only applicable to that specific country in most cases. Global agencies on the other hand, tend to implement central campaign management teams which already have expertise working with marketers in different countries, experiences and knowledge that can be transferred and applied from one country to the next. Ideally, the global agency has an office located close to the client for matters of planning and contact, but can use a central hub for campaign management and execution. The implementation of a central hub creates the opportunity to draw on expertise from the local market and share experiences all in one place. This helps to build a strong international network but is also valuable for using local expertise on a global scale.

2.2 The Data Aspect

As with inventory, the number of third party data suppliers connected to the programmatic advertising platform is growing. Third party data is largely composed of demographic information (e.g. age, sex) that can be used during a campaign. The easy access to data such as demographics provided by programmatic advertising is an advantage for smaller advertisers.

Nevertheless, in many markets there are problems with the availability of third party data. High-quality data, which enables accurate targeting, is rare in a market. The volume that is required is simply non-existent. For this reason, the modelling of targeting criteria on the basis of existing data has become standard in some markets and as a result the quality of data has become a decisive factor. High-quality data is collected and shaped, "statistical twins" via look-alike targeting are formed and targeting profiles are thus defined.

Some targeting segments from data management platforms based on high-quality data sources can be used to supplement the often limited targeting opportunities in programmatic advertising platforms via third party suppliers.

Individual segments and targeting criteria that differ from other suppliers, and that are not accessible to everyone through programmatic advertising are essential in order to remaining competitive.

This is the reason for the rise of data management platforms (DMP). The three key advantages are:

1. Expansion of the programmatic advertising data pool through various data sources
   Connecting third party data to programmatic advertising platforms results in different targeting criteria. However, the range of targeting criteria is limited. To guarantee the client can achieve the expected reach, targeting suppliers use a number of different methods. For example, Xaxis expands its programmatic

advertising data pool through various offline and online data sources in Turbine, which is its proprietary data management platform, and this serves as the basis for targeting segments. The result is a combination of first, second and third party data, which is continuously collected, structured and dynamically processed for implementation within the data management platform.

2. Transparent quality management
With the data management platform, which is connected to the programmatic advertising platform, the supplier can clearly show what is implemented for whom based on what data. This independence provides transparency regarding the origin and processing of data and ensures an uncompromised data handling system.
3. Diversity of segments that can be offered
Data sources define which targeting options can be offered. The large selection of data provides opportunities to offer customised segments that go beyond traditional metrics and pre-defined target groups. For example, customised target groups such as lifestyle segments or buyer classifications can be established and addressed. These customised segments are essential in order to stay one step ahead of competitors and offer added value.

On an international level, commerce for data can be centralised. This usually makes the purchase of data more cost-effective. The quality and volume of data vary from country to country and should first of all be evaluated. Collection methods often vary too. From a technical point of view, central and transnational data management is an effective way of making the same data available in all countries. If the treatment processes and algorithms that access the data are standardised, new data sources can be connected to the data management platform with minimal effort, or old ones exchanged. In an ideal case, the data-based profiles thus have a global presence simultaneously and can be used across all markets using programmatic advertising.

3   Overview from an Advertiser's Standpoint

The above insight has shown us how central, transnational campaign management opens up completely new opportunities. However, the question remains in which case advertisers should choose central campaign management across countries versus local campaign management. Table 1 shows a comparison demonstrating when central campaign management may be preferable to local campaign management and vice versa.

This contrast does not mean that advertisers need to choose between the two forms. The choice between central or local campaign management is defined by the campaign's objective. Depending on the task at hand, both forms may be used by the same advertiser. This is especially useful for larger advertisers with a marketing division (e.g. umbrella brand campaigns) that is managed by a central team. In addition, teams from individual countries oversee other local marketing areas (e.g. single brand campaigns). For such requirements, advertisers will ideally work with agencies that can satisfy both disciplines according to the advertiser's needs.

Table 1 Central versus local campaign management (own research)

	Central campaign management across countries	Local campaign management
Marketing structure	Strong, central marketing organisation; transnational planning	Decentralised marketing organisation; planning for individual markets
Campaign set-up	Standardised campaign set-up across multiple markets; Little variation and market specification	Complex campaign development; A lot of country-specific variations
Campaign KPIs	Performance can be measured uniformly using the advertiser's websites (e.g. online shops) across different countries	Performance measured differently for each country using various advertiser websites
Advertising formats	Standardised; Can be used across all countries	Very varied for each country
Quality management of inventory	Agency has standardised black/white lists throughout the world; Brand safety tools	Individual black/white lists for each market; Need for strong local expertise
Targeting criteria	Standardised; Can be used across all countries	Individual targeting profile based on data available in each country
Reporting	Standard KPIs; Can be used across all countries	KPIs for individual markets; Can only be used locally

4 Conclusion

This insight into borderless media management shows us how, as a result of programmatic advertising, the market for inventory and data has become a large international department store. Global campaign planning has caught on across the entire market and will continue to grow with the establishment of programmatic advertising. On the one hand, it offers new opportunities for planning campaigns that span national borders. On the other hand, competition is intensifying. Market participants should be aware of these effects and ask themselves the following questions to address the situation:

- From an advertiser's perspective: Should I choose local or central campaign management?
- From an agency's perspective: How can I offer clients a global service?
- From a publisher's perspective: How can I offer my inventory on a global scale using programmatic advertising?

Although international campaign management is a common process, boundaries are set. Local expertise continues to be absolutely essential. In future, the trade in digital advertising space will be conducted in a similar manner to normal trade. Just as in other walks of lifebuying behaviour varies according to our requirement –

sometimes we buy locally from around the corner, sometimes online from a domestic online shop and sometimes online from overseas. It's getting the right mix that counts.

Bibliography

AppNexus. (2014). Internal analysis on buyers and sellers for the global market.
IDC. (2013, November). Worldwide and U.S. real-time bidding 2013–2017 forecast. http://www.idc.com/getdoc.jsp?containerId=244194
US Programmatic Ad Spend Tops 10 Billion This Year, to Double by 2016, emarketer (2014, October 16). http://www.emarketer.com/Article/US-Programmatic-Ad-Spend-Tops-10-Billion-This-Year-Double-by-2016/1011312
Xaxis. (2013). Internal sales report of Xaxis EMEA markets.

Rosa Markarian is Light Reaction's Vice President, Global Business Operations. She is responsible for business metrics, product enablement, training, community and new market support for Light Reaction, Xaxis's mobile performance marketing company. Rosa has a wealth of experience in the digital marketing space from her work at Xaxis where she was instrumental in building the company's business and products in the German and European markets. Prior to Xaxis, she worked in strategic business development, leading projects with a digital focus in the mail order sector at the Otto Group. She started her career as a pioneer in the field of targeting, having developed products for Germany's first targeting technology at United Internet Media. A leading expert, Rosa Markarian is a regular speaker at industry events and a published author, contributing articles to ad trades on online advertising.

Dr. Aee-Ni Park is Global Director of Platform Product Manager of Xaxis. She is responsible for the development of new products and algorithms of the Data Management Platform (DMP). In her previous role she managed the product team in Germany. Before joining Xaxis, Aee-Ni worked for the global VivaKi network focusing on strategic technology and data mining.

Having completed a PhD in computer science, Aee-Ni holds a strong technical and research background.

Dr. Mark Grether is Global COO of Xaxis, a digital media company servicing over 2,700 clients in 40+ markets across North America, Europe, Asia Pacific and Latin America. Xaxis programmatically connects advertisers and publishers to audiences across all addressable media channels through the expert use of anonymous consumer data, advertising technology and media relationships. Advertisers working with Xaxis achieve higher ROI from digital marketing campaigns. Publishers deliver relevant content and advertising to new and valuable audiences. As Global COO, Mark is responsible for global operations, including product development, media and data strategy, technology development and integration as well as internal systems. In this capacity, Mark is responsible for defining, creating and delivering best-in-class product and technology offerings globally. Prior to founding Xaxis, Mark worked as the Global Development Director of GroupM. As the founder of GroupM's global audience initiative, he was instrumental in setting up GroupM's digital trading desk across ten countries. He was in charge of managing all processes in the value chain, from defining trading strategies, product definition and technology design through sales. He also oversaw the change management process within the GroupM agencies. Mark has a Master of Business Administration from the University of Mannheim in Germany, a Master of Arts in Marketing from the University of Florida, and a PhD from the University of Mannheim. Mark's work has appeared in more than 50 publications and he has been a featured speaker at conferences across the world.

Programmatic Disruption for Premium Publishers

Holm Münstermann and Peter Würtenberger

For publishers of journalistic websites who offer contextual advertising, the rise of programmatic advertising represents a particularly big challenge: They risk being demoted to mere suppliers of ad space, as data points for determining target groups can be purchased separately. Publishers can only overcome this challenge by taking the plunge into programmatic, data-based selling – and buying.

1 Origin and Characteristics of Premium Publishers

There's hardly a term used more often in marketing than the adjective "premium." There are "premium reaches," "premium inventory," "premium marketers," "premium sales" – all of which are usually used to elevate things that aren't premium and are usually no different from the products and services of others.

Publishers used to be "premium" by nature – as publishers of newspapers and magazines, or distributors of journalistic content. But today, at a time when any high-school student can produce, publish and operate their own website for just a few dollars a month, the term "premium" is used to distinguish between sites featuring journalistic content produced by professionals, and those that are written by amateurs or where the focus isn't even on editorial content. As a result, professional content sites can be considered premium, while the term is used less and less for sites made up of service-generated (e.g., mailings) or user-generated content.

H. Münstermann (✉)
Axel Springer Media Impact, Axel-Springer-Str. 65, 10888 Berlin, Germany
e-mail: holm.muenstermann@axelspringer.de

P. Würtenberger
Axel Springer SE, Axel-Springer-Str. 65, 10888 Berlin, Germany
e-mail: peter.wuertenberger@axelspringer.de

© Springer International Publishing Switzerland 2016
O. Busch (ed.), *Programmatic Advertising*, Management for Professionals,
DOI 10.1007/978-3-319-25023-6_3

Even if the difference between premium and non-premium publishers is not entirely the same because, from a marketing perspective, every website would naturally like to be considered premium, it is indisputable that the web offerings of magazines and newspapers are premium sites – and by default, their owners are premium publishers. Publishing houses and companies therefore make up the core of premium publishers.

And it was also publishing houses that transferred their marketing principles from print to online and are still successful with this today. For example, ad space on the homepage of Germany's largest news site, Bild.de, is sold at a cost per day – just like in traditional print papers. And it was also publishing companies that applied their principle of placing advertising around journalistic content to marketing their online sites: Special, theme-based ad placements have much higher rates than campaigns where an ad server determines where on a site or network a campaign is inserted next, based on availability.

This approach to selling ad space, taken over from newspapers and magazines, quickly became known on online portals as "contextual advertising." Because in the world of print advertising, which is hard to quantify and measure, the context is the only targeting criteria, the only data point, which offers an advertiser some orientation: For example, if there's an article about the common cold, it's likely that it will be read by people who currently have a cold or at least suffer from one frequently. The waste circulation when advertising cold drugs is therefore significantly lower when the ad is placed next to or below an article like this than when it's placed next to a sports article, for example. The journalistic article reduces waste circulation by offering an appropriate context – because the context is suitable for coming to a conclusion about the person who's reading the article and therefore looking at the ad next to it.

The principle of contextual marketing is why, in online marketing, journalistic sites achieve a higher cost-per-mille (CPM) than mailing websites, for instance: The article determines the reader and, therefore, the user, and as a result is little more than a targeting measure. Additionally, people who read articles typically spend longer than average on a single page impression. Therefore, besides targeting, the ad's increased visibility causes a CPM markup in the contextual marketer's rates.

Premium publishers in online marketing are therefore publishers of digital, journalistic content, which they can generally also market at premium prices – because they can offer professional environments that reduce waste circulation for ads.

2   Existing Disruptions for Premium Publishers in Digitalization

In the late 1990s, the publishing industry began putting its content online. What's now seen as a major error went without question back then: The articles were made available online, completely of charge. They were to be financed via advertising. Therefore, advertising also had to compensate for this second, lost income stream. The misnomer "cannibalization" was increasingly used: The same content that generated both sales revenues and very good advertising revenues in print was put online – where sales revenues weren't the norm and advertising income hadn't yet been developed. But the exodus of readers from print to online couldn't be stopped, and titles that didn't follow suit risked losing their readership completely.

The transfer of content to the Internet therefore represented the first disruption for premium publishers. This disruption still hasn't been overcome to this day: In the meantime, the majority of the big titles have reached as many unique users with their online versions as readers with their print editions – but in most cases there is still a huge gap between advertising revenues per user and per reader. In previous years, annual increases in online advertising revenues meant this was no cause for concern; it only seemed a matter of time until the gap was closed. But for premium publishers, advertising on desktop sites (as opposed to mobile sites) is by far the biggest source of digital advertising revenues. And even this will decrease in 2015 – at least in the USA, according to eMarketer.com. This shows that the race to catch up has come to an end: Display revenues per user will never reach the levels of advertising revenues per print reader.

What's even more tragic about this development is that, not only were the levels of print advertising revenues missed, but that most publishing companies still don't make any online sales to compensate for the falling sales of print versions. Without paid content, journalism will be stripped of its financing in the long term.

However, a publishing house must go where its readers go. And so began the second digital disruption for premium publishers worldwide, which started with the introduction of smartphones at the latest: Users are moving from stationary to mobile Internet access. In 2014, for the first time, more users accessed the Internet on their smartphones than their desktops (see Fig. 1). But for sales revenues, this is no longer a disruption – it's an opportunity: Apps are becoming the norm in mobile Internet access, and these apps are procured via a digital kiosk with a payment function – an app store. However, marketers targeting sales revenues are already facing the next disruption, without having overcome the first: It was difficult enough to transfer large, eye-catching 1/1 ad pages from newspapers and magazines to desktops, but it's almost impossible on an even smaller screen. Ads can no longer be placed next to articles and therefore benefit from the context – they have to be placed above or below the content, or cover it completely.

The best chance for contextual marketing on mobile devices is to place the ad *inside* or at least *between* the content and make it look like content. The buzzword "native advertising" is used for this, and has become a nightmare for compliance officers at publishing houses. While Facebook uses the same text and same layout, just with a small note that says "sponsored" to let users know that a post represents paid advertising, publishers must ensure the strict, clear separation of content and advertising, to prevent their independent journalism from being corrupted by the advertising industry. This is an important principle, but one that makes the transition of readers to mobile devices even more painful – because the most significant form of advertising there is inaccessible to premium publishers.

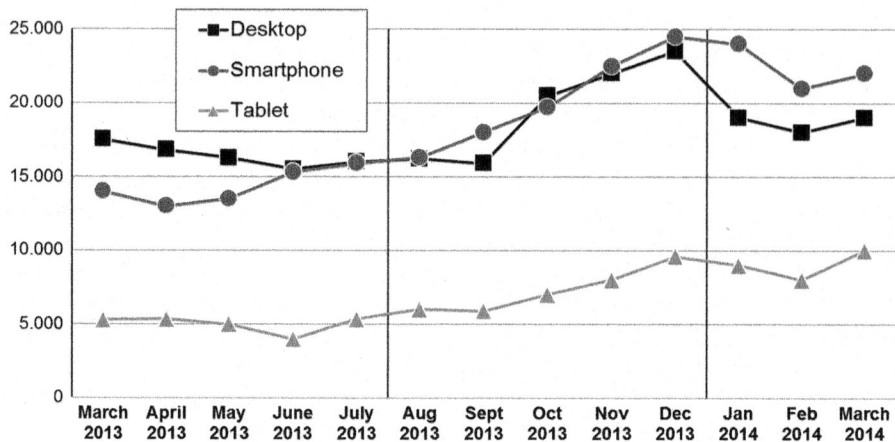

Fig. 1 Internet usage, minutes per month and user (Source: comSource media metrix and mobile metrix, U.S., Feb-2013–Mar-2014)

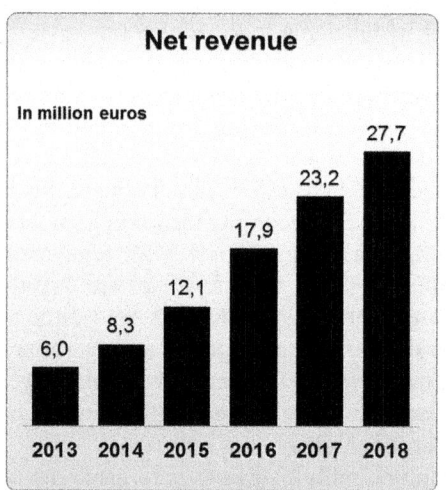
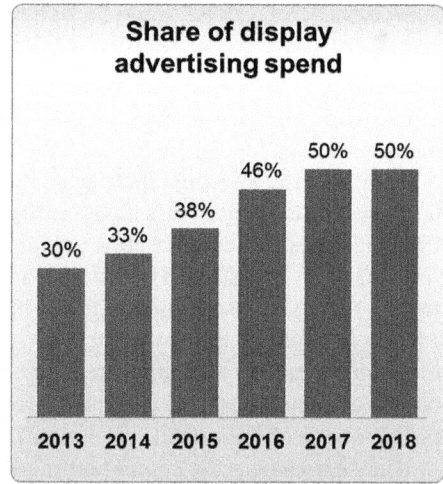

Fig. 2 Growth of programmatic display revenues (Source: BI Intelligence Estimates, Magna Global, IDC)

And while the first disruption hasn't yet been overcome and the full extent of the second hasn't yet been determined, the triumphant onward march of programmatic advertising (see Fig. 2) represents a third disruption for contextual marketing – one that challenges premium publishers existentially in terms of their self-understanding.

3 The Significance of Programmatic Advertising for Premium Publishers

Programmatic advertising is essentially nothing more than the long overdue automation of buying and insertion processes in digital advertising (see Fig. 3). Even today, large online marketing campaigns are still ordered with a signature and stamp on a quote, and aren't inserted until the corresponding fax comes through from the agency or buyer. It's high time this process was modernized. Now the advertiser or its agency can deposit the insertion order plus ads on a demand-side platform ("DSP"), which interacts with the marketer's supply-side platform and puts the campaign on its site.

One particular form of programmatic advertising is Real-Time-Bidding (RTB), where auctions ensure that the maximum a bidder is willing to pay is exhausted for every ad impression, in milliseconds. This only works to the publisher's advantage if multiple bids are made for the same impression. But it works: In fact, automated buying and insertion processes can be used for even small campaigns that don't meet the minimum order volumes of media agencies and marketers in direct sales. Without incurring any sales expenses, a publisher has access to additional demand, which, thanks to the auction mechanism, boosts CPMs in the non-guaranteed delivery segment. With Real-Time-Bidding, publishers can increase the rates and therefore revenues for their long-tail inventory. For all inventory that until now has been sold at fixed CPMs, but without guaranteed delivery, Real-Time-Bidding allows the true price customers are willing to pay to be calculated and exploited – which works to the *advantage* of the premium publisher.

The challenging disruption for premium publishers comes from the fact that it's now possible for advertising customers:

(a) only to bid for ad impressions that match their target group, and
(b) to select their target group without considering the contextual environment of the publisher whose ad space they are using.

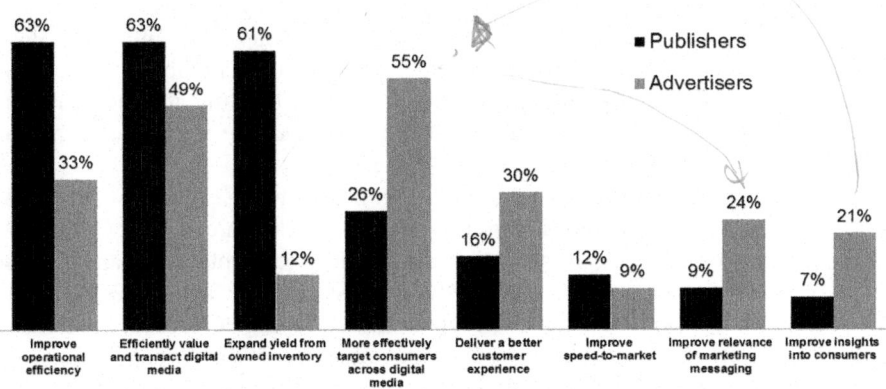

Fig. 3 Drivers of programmatic adoption (Source: Winterberry Group, IAB 2013)

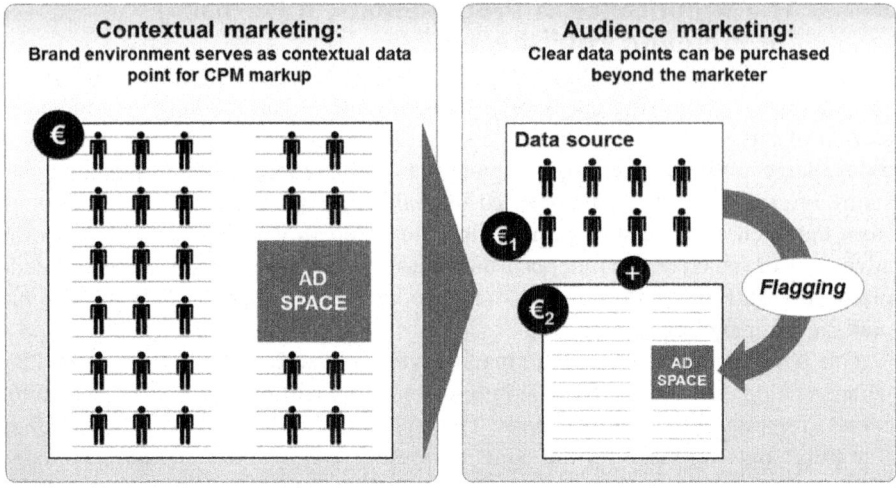

Fig. 4 Change in the way display advertising is sourced

Programmatic media buying platforms allow for something akin to a game of table tennis with user databases, which, in the milliseconds between the ad server's ad call and actual insertion of the ad, lets an advertiser know whether the user behind an impression belongs to its target group segment. It's these user databases, or data management platforms, which are taking over a part of the added value offered by contextual marketing – namely, target group data. This leaves only the second part for the premium publisher, the provision of advertising space. These two interwoven elements – space and target group data – which previously determined rates of digital contextual campaigns in the "old world," can now be sold separately (see Fig. 4).

Here's an example: An editorial portal that reports on computer innovations and performs complex comparison tests of hardware for its readers used to be able to rely on high revenues from marketing. This is because advertising clients such as computer manufacturers, who want to reach a target group that's into technology and, ideally, is planning to buy a new device, could only minimize their waste circulation and achieve the best click-through rates by booking ad space on a site like this. That's why a much higher rate could be charged for the same ad format on this site than on general or non-editorial sites.

Thanks to programmatic advertising, the computer advertiser can now source the data points of a price comparison portal and pay a revenue share or fixed CPM to the data supplier. For example, the advertiser uses anonymous cookies to mark all those users who searched for computer hardware costs in the last 4 weeks. In all likelihood, these users are currently considering buying a new computer and looking for relevant advice. Using a DSP, the advertiser can now bid for ad space – but only if the user who's just visiting the website corresponds to the cookies of the selected user segments on the data management platform. This means the advertiser can target its audience in a much more direct way than by buying ad space on the editorial computer site. That's because this site's readers may simply be interested in computer innovations, without having a concrete intention to make a purchase.

Similar examples can be seen in all online environments, which, until now, have been able to achieve high CPMs because of their target-group specific content: An auto website, whose core clients will in future bid for advertising based on the user segments of a new car configuration site, and not necessarily on the auto site. The fashion website, whose core clients will in future employ semantic targeting to find the right advertising environments, but will procure them for the price of nontransparent run-of-network campaigns instead of transparent contextual campaigns on the fashion site. The travel website, whose regular advertisers will now procure their ad space based on the user segments of a travel booking engine where target users are recognized – and not necessarily on the editorial travel site.

For the first time in the marketing history of premium publishers, the value of the contextual environment as a data point has been thrown into doubt – because there is now an alternative for buying ad space based on user data. With programmatic advertising, data points can be purchased independently of the space and used as a pre-targeting. This is why programmatic advertising represents the next, and possibly the most existential, marketing disruption for premium publishers.

4 The Consequences

The consequences of the programmatic disruption for premium publishers are easy to foresee – but their scale isn't. We want to focus on the following three consequences:

(a) The demise of premium CPMs
(b) The disappearance of the mid-tail
(c) New competitors

4.1 The Demise of Premium CPMs

As highlighted in the previous chapter, the data point as a key cost benefit of advertising with premium publishers can now be found elsewhere. An advertiser can determine which users should be reached on a website using a good DSP, in return for an additional charge. Because this criterion is based on the aggregation and segmentation of various behavioral or CRM data, in some circumstances it may even be superior to profiling users by visiting individual sites. But the additional cost that premium publishers charge for ad space on their sites isn't usually a percentage of the cost of the same space on non-premium sites. It's usually a multiple of that.

Advertisers will continue to use the familiar environments of premium publishers, if only for reasons of "brand safety." But they will no longer use them as targeting criteria, rather as brand-strengthening premium environments – like a perfume that's considered higher quality in the cosmetics department than at a discount retailer. But premium publishers will only be able to maintain their existing rates, if they add further targeting measures to their impressions, as described in the following chapter. Otherwise, as soon as programmatic advertising takes more than one-third of the online display market in their country, they will see a fall in their average CPMs for run-of-site or channel usage.

4.2 The Disappearance of the Mid-tail

The mid-tail segment is the bread and butter business of premium publishers. While the top homepages are sold exclusively and as a rare commodity, and the long-tail as non-guaranteed, run-of-network space that offers basic circulation with low CPMs, the mid-tail represents high-margin, second-best placements. This is space where run-of-network campaigns are also featured, but where delivery and placement can be guaranteed. It does not include the very best ad space such as on top homepages. But it is significantly more expensive the long-tail segment.

Programmatic advertising means that the mid-tail has new competition from below (see Fig. 5): Run-of-network campaigns can suddenly be managed fairly precisely – not only by the publisher, who may demand an additional charge for it, but also by a good audience marketer via a DSP. The client defines the most important terms for their product, and then the system crawls the Internet for sites featuring these terms, adds new, appropriate terms to a "tag cloud" and then continues to browse for sites which best fit this tag cloud. This form of "semantic

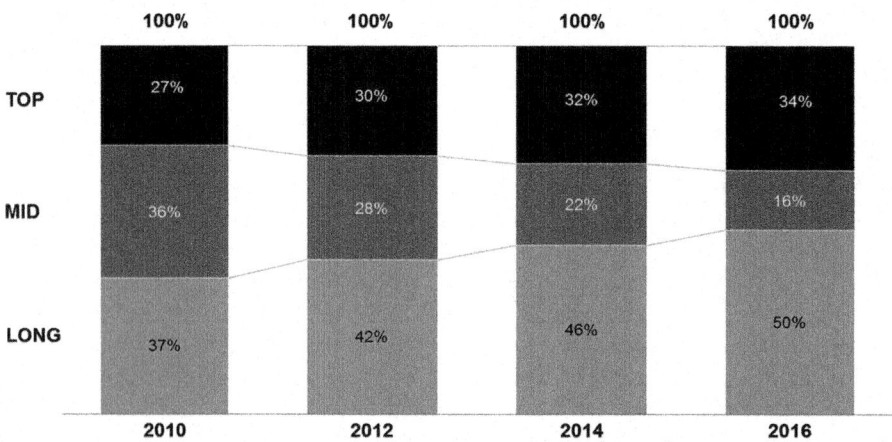

Fig. 5 Forecast revenue shares of inventory segments in the US display advertising market (Source: Credit Suisse Study "Web 2.012" 2012)

targeting" helps to narrow down the environments where the advertiser wants its campaign to land. But the CPM, which is paid to the publisher for this, remains that of a long-tail campaign. If a publisher fails to enhance its mid-tail segment with its own data, this segment will be pushed out by the "external" long-tail, i.e., by external data sources that the advertiser can integrate. However, even with programmatic advertising taking a much bigger share of the market, the top placements will be able to retain their premium prices, and as a rare commodity may even benefit from the auction mechanism of Real-Time-Bidding.

4.3 New Competitors

It was never an advertiser's primary goal to place its ads in a specific environment. Placement was always intended to help it reach its target group. But while contextual marketers at premium publishers primarily offer the website that they are representing, audience marketers take the opposite approach: They inquire about the client's target group first.

For many years these audience marketers were mere ad networks – selling unfamiliar ad spaces, which they aggregated and repackaged as verticals of complementary environments. Nowadays, programmatic advertising means anyone can allocate his verticals to environments – you don't need an ad network anymore. All of which means audience marketers, who enable a campaign to be run based on exclusive, hard – and not extrapolated – target group data, have been pushed into the foreground. Almost every large owner of digital customer data is becoming a marketer by offering marketing products based on its data segments. The largest of these are Amazon, eBay, Apple, Google and Facebook, which are all sitting on huge volumes of customer data. But advertiser data like that collected by Criteo can also create large marketers. Media agencies have also created trading desks, to collect data and use it for their own marketing purposes. Otto Group, the mail order multi-national based in Germany, has just established an advertising sales house – not because of its expertise in marketing, but because it owns so much great data. Almost all companies that own data have become potential competitors for the sales houses of premium publishers, battling for the same advertising budgets.

5 Potential Approaches for Premium Publishers

Premium publishers can't afford to be reduced in their roles as suppliers of advertising space. CPMs like non-premium publishers, the disappearance of the high-margin mid-tail segment, and smaller market shares due to new competitors will cause the contribution margins for financing editorial offices to shrink and, with that, the foundation for professional journalism. Publishers and their marketers cannot sit by and watch this development unfold – they must act on it.

The following measures are available, as a minimum:

5.1 Profiling Their Own Users

Most premium publishers have already set up their own data management tags. However, these are rarely used to systematically establish new segments because the overwhelming opinion is that a data-driven targeting offering could usurp the individual brand environment and therefore support the flow of advertising budgets into audience marketing. This is true to some extent, but as long as the revenue remains with the same marketer, it should still be promoted – as the alternative will happen anyway.

Data segments could be formed based on advertisers' interests. Every industry has its own specific target groups that it wants to reach. Asking which of the premium publishers' channels and products users of these target groups would use and read helps to generate and name the segments with the aid of a data management platform. The underlying principle still follows the combinations of themes or verticals that are already offered in contextual marketing, but can be technically expanded by setting up additional rules or conditions (e.g., "user visited site A and site B in the last two weeks"). And the price point can be calculated dynamically, with the publisher programmatically allowing advertisers to bid on its inventory with the new segments as an additional targeting charge, or making a bid on it itself to prevent higher-priced campaigns from being squeezed out.

5.2 Generating Registration Data

For years now premium publishers have neglected the concept of user logins. Of course, beyond paid content, there are also few reasons why a user should register with a news site. But registration data could become a matter of survival for premium publishers. If, one day, web browser creators or even European legislators question the use of cookies, only those companies with registration data available to them will be able to charge more than basic ad rates – with the four big US players Facebook, Amazon, Google and Apple leading the way. These four will also dominate cross-device marketing because cross-device tracking is only 100 % certain with logins.

Premium publishers will therefore have to find ways to convince their users to register and then remain logged in on all devices. To do this, they need to create added value – for example, through competitions, free services, rewards systems or providing arguments against registration such as restricted access to content.

Here the exodus of users to mobile devices represents an opportunity – because people are used to registering for apps. However, there is still the challenge of persuading users to download an app in the first place and to continue to use it frequently.

5.3 Buying External Data

Where a publisher's data isn't comprehensive enough to provide easily marketable segments, e.g., a lack of reliable sociodemographic data (gender, age, residence), these segments can be bought and added to the publisher's own DMP segments. For example, a female user is known to be interested in cars because she's read several articles about them in the publisher's own portfolio – but it's only confirmed that she's female by a third-party data provider. It is, however, crucial to add an exclusive element to these segments – even if it's simply the user's visit to the publisher's own sites.

5.4 Buying External Inventory

In general, the more complex and precise a data segment, the smaller its reach. If a segment is only to be used in the publisher's own portfolio as part of a campaign running for a limited time, the reach is further reduced to the crossover between the limited data segment and the user's visits to the publisher's site during the campaign period.

However, creating segments on a DMP allows the publisher to recover these data points from external inventory and source them as needed, at any time. Every marketer should make use of this opportunity, without losing sight of what is most important – its clients. But it's also in the interest of advertisers to extend their reach through use of their own data points; as well as looking for increasingly precise targeting, they want a minimum reach. Advertisers and publishers can work more closely in this area: Crossovers between the advertiser's customers on their own platforms and a publisher's users are analyzed and checked for patterns. The same patterns can then be searched for among the publisher's users, because they suggest a high degree of affinity with the advertiser's offering. Finally, these users can be marked, searched for on external sites (beyond the publisher's sites), and furnished with advertising purchased by the publisher on behalf of the advertiser.

5.5 Charging More for Premium Space via Brand Safety Scores

The online environments of premium publishers remain premium. Even if they are not sufficient enough as sole targeting criteria in future, they remain a "hygiene factor," bringing added value for advertisers. In contrast to publishers' current rates, there could be a basic price per ad format – similar to a pizza base. This could be enhanced with a selection of various "toppings," i.e., data segments, as well as a choice of online environments of different quality. For measuring the quality of an environment and calculating the additional charges, a "brand safety score" could be produced that ranks environments based on how high quality they are.

However, this approach would only be recommended if audience marketing really were squeezing out the majority of contextual marketing. In this case, it would be sensible to allow the advertising client to first choose the target group (= user segments), then the format, and finally the quality of the environment based on the brand safety score.

6   Conclusion

Right now, marketers of premium publishers are in the middle of their third digital disruption. While advertisers are benefiting from programmatic advertising, premium publishers must take action to ensure that they are not degraded to mere suppliers of ad space and do not have to lose large chunks of their contribution margins. To do this, premium publishers need to learn from the very audience marketers who represent their new competitors. They must make their data available for smart segmenting and enhancement, and buy external data and inventory. Only then can they make the most of their professional content sites and long-lasting, trustful client relationships to ultimately successfully secure advertising budgets and premium CPMs.

Holm Münstermann is General Manager for New Media Business at Axel Springer Media Impact and, in this role, has introduced data-based marketing and programmatic advertising at Axel Springer's sales house. Prior to this, he led digital marketing at Axel Springer Media Impact, was Executive VP at the digital agency Pixelpark AG, and was also Director of International Operations at the Bertelsmann e-commerce subsidiary bol.com. Münstermann holds a degree in Business Administration from Cologne University and has a CEMS Master degree in International Management from Cologne University and HEC Paris.

Peter Würtenberger is Executive Vice President Corporate Development of Axel Springer, Germany. Until April 2015, he was Chief Marketing Officer at Axel Springer, heading Axel Springer's central sales house – Axel Springer Media Impact. Previously, he was the Managing Director of Newspaper Group Berlin, which publishes Welt, Welt am Sonntag, and Berliner Morgenpost. Würtenberger served as Chief Executive Officer of Bild. T-Online Berlin, Managing Director of Yahoo! Germany, Chief Financial Officer of BMG Berlin Musik, and CFO of BMG Classics Europe, Schlumberger Oilfield Services, Amsterdam, the Netherlands and Hanover, Germany. Würtenberger holds a degree in Business Management from Cologne University and also completed the Advanced Management Program at Harvard Business School.

Perspectives of Programmatic Advertising

Jürgen Seitz and Steffen Zorn

Despite its bright future, programmatic advertising still faces significant hurdles, such as a lack of best practices beyond remarketing and premium inventory, the problem of cross-device adaptability and a lack of subject experts. In addition, there are risks such as the increasing use of ad-blocking software, potential restrictive privacy regulations and an increasing focus on CMOs' own media channels and content marketing approaches. Similar to the Internet's development, the authors see an initial hype, a subsequent disillusionment followed by a continuous rise of programmatic advertising. This implies a significant change in the value chain. Established global suppliers are particularly well positioned, but truly innovative start-up companies also have an opportunity to succeed in this environment.

To fully exploit the programmatic advertising potential, the focus of data-driven advertising and marketing needs to be extended to channels & formats beyond display advertising such as video, social media, mobile and native advertising. For content creation and distribution, programmatic approaches are also of growing importance. With an increasing use of programmatic systems, the user becomes more important; the mapping of customer journeys across channels becomes elementary. The communication context is becoming more important than the channel. The recipient uses his mobile at home and while commuting, but his needs at home differ significantly from the needs while being on the road.

Due to explosive growth in available data, increasingly affordable and high-performance databases, and growing in-house analysis capabilities, knowledge of relevant data points and user groups will significantly increase. Instead of

---

J. Seitz (✉)
Hochschule der Medien (HDM) Stuttgart, Nobelstraße 10, 7569 Stuttgart, Germany
e-mail: juergenseitz@web.de

S. Zorn
Swinburne University of Technology, 6 Hicks St, Karrinyup, WA 6018, Australia
e-mail: steffenzorn@iinet.net.au

individual campaigns, continuous and targeted marketing programs are increasingly popular.

The imminent problem of monetizing mobile traffic is a major driver for growing data-driven marketing performance. Media publishers will either solve this problem through an intelligent combination of data and content or they will perish. But mobile web technology is also a huge opportunity. The "Web of things" drives programmatic advertising to new dimensions. The combination of real-world user data and online data will offer new opportunities. Despite all technical enthusiasm, the tension between human marketing experience and machine learning remains. People do business with people; technology has a supporting function. Programmatic meets real life.

1 Quo Vadis Programmatic Advertising?

When speaking with digital marketing experts about programmatic advertising, even very experienced digital managers admit that they can hardly keep up with the speed of development and the variety of options. In a very short time, an extremely dynamic and complex ecosystem for technology and data-driven online advertising has emerged, which is very difficult to understand.

Experts agree on the overall forecast. IDC, JMP, JP Morgan or eMarketer – all see strong growth over the coming years. Unfortunately, these predictions have a similar accuracy as long-term weather forecasts. They give overall guidance, and forecasting methods improve; however, due to the remaining uncertainty, there is limited value for marketing decision-makers. For strategy and concept development, predictions require the right accuracy to be of practical value.

In which part of the market are the highest growth potentials? Which platform will thrive? What is the share of the media volume and in which industry? When will this be the case? Are programmatic advertising initiatives profitable? To make things worse, behind statements about future programmatic advertising development there is often a tangible corporate agenda and probably just as often a personal agenda of an individual manager. Which senior marketing or agency manager wants to be in a completely unknown territory without any need? Similarly, a technologist in an ad-technology start-up will hardly admit that a marketing manager predicts trends better "with the right feeling," or that simply replacing a creative has outperformed the technologist's algorithm. This is an ongoing conflict in programmatic advertising.

2 Programmatic Advertising in the Hype Cycle

Like every other new technology programmatic advertising follows a traditional hype cycle. After a slow start followed by a steep increase in awareness, expectations and enthusiasm, a sudden crash happens. The euphoria fades faster than it came. The decisive factor is how fast the following slower but more sustainable revival is progressing and which market participants establish themselves with the right timing.

Experienced online marketers can recognize similarities in the development of the Internet and now programmatic advertising. Exaggerations leading to the first Dot.com bubble around 2000 resulted in a painful "ice age" but could not stop the development of the Internet. Many business models that were highly celebrated at first but slashed as irrational after the stock market crash, work very well in the meantime. Look at Amazon or online pet food sales figures.

The answer to the question where programmatic advertising stands in the Hype Cycle depends on your personal point of view. Some are bullish, others don't believe the hype. Wherever you stand, one thing is clear: Uncertainty regarding current conditions and the future development of a sector is usually an inhibiting factor for investments, however, positive opinions regarding programmatic advertising are the majority.

3 High Willingness to Invest Despite a Remaining Uncertainty

Questioned about global e-commerce market expectations Internet entrepreneur Oliver Samwer answered: "It's big enough, it's just bloody big enough". It seems that CMOs of companies and agency leaders think alike about the programmatic advertising market. All of them invest in programmatic advertising as part of the ever-increasing IT investment in marketing. Only recently it was predicted that marketing departments would soon have the largest IT budgets in many companies. Furthermore, they create management roles like Chief Data Officers and put them into the center of their efforts. Oliver Samwer has not only proven that overwhelming optimism can lead to huge success, with his team at Zalando (a Zappos clone), he has also shown that it is worthwhile to invest in programmatic advertising. One of the main factors contributing to their worldwide "Made in Germany" success story in e-commerce is the extensive use of data-driven advertising. Remarketing worked especially well for them. Criteo, one of the European success stories in online advertising, developed a very successful business on this approach and is now rapidly moving forward in other fields of programmatic advertising like prospecting.

4 Fundamental Changes in the Ecosystem of Advertising

The global rise of programmatic advertising concepts and related software is – as often in the Internet area – primarily an American success story. The "bold vision" of programmatic bidding, auctioning display advertising inventory like Google's keyword driven ads, is the cornerstone for the present, broader vision of programmatic marketing. It is no longer "just" auctioning online inventory, but rather a complete consumer focus, the aggregation of all possible data available and the analysis of a consumer's current needs and basic preferences. Furthermore, more and more companies are taking a closer look at the decision process of their customers. Instead of using simple sales funnels, they analyze the journey of the customer, identify the major touchpoints and evaluate the contribution of each touchpoint to the generation of sales. Based on this concept, the contact opportunity with the consumer is valued and sold in programmatic platforms. Such a radical consumer focus must inevitably be accompanied by a fundamental change in the value chain.

The cornerstone of programmatic advertising has been established in the United States, and US global providers drive the operative conversion towards people-oriented programmatic advertising. Besides numerous programmatic advertising niche companies, mainly US-based service providers, Google, Amazon and Facebook scale their online advertising revenue globally through data and programmatic automation. The major media agency networks also use their market power for the development of global programmatic advertising models such as Xaxis.

The strong marketing performance of Google's Display Network (GDN) or Facebook Advertising results from massive data ownership, own unique identifiers, consistent data-driven Programmatic optimization, a high usage intensity of their online offerings and aggressive investment strategies. This new ecosystem will result in a significant change in the value chain. Purchasing power, planning know-how, and connecting advertisers and media as previous core-value drivers of media agencies are complemented and partially replaced by data ownership, analytical knowledge and programmatic advertising infrastructure.

In contrast to technological possibilities of unbundling media offerings (data, placement etc), major market participants seem to integrate several stages of the value chain into an end-to-end offering (full stack). Consequently, the resulting simplification seems to attract a lot of budget. Therefore the percentage shift in the value chain will be accompanied by a shift in market shares to channels and offers with large advertising values. Low value or small sized market players in the value chain are increasingly irrelevant. This does not mean that entrepreneurial companies will definitely lose the battle, but they have to provide true value beyond pure sales, and they have to grow beyond a certain threshold. In a landscape that is becoming more complex due to changed media consumption, marketers and agencies are searching for simplicity and scale from service providers and media owners. Yet too many media companies act inexplicably passive and react slowly. These companies will hardly play a relevant role. Other media companies and media agencies will transform – voluntarily or involuntarily – into conglomerates. They struggle to compete with global platforms and emerging new online pure players. Consequently, they seek their salvation in acquisitions within fields such as e-business, e-commerce and in the field of specialized media services. Disadvantages of these valid strategies include often significantly lower margins in these newly acquired business areas, the pressure to become more efficient in the remaining publishing area, and simultaneously low quality journalism. The resulting consequences for journalism will lead to heated debates in society and hopefully an increased willingness to pay for premium content. Whether programmatic advertising actually delivers the promised advertising revenue increases for content providers remains questionable, despite its current success.

5   High but Manageable Hurdles

Despite all excitement or fear – depending on the point of view and assessment of the Hype-Cycle stage – regarding programmatic advertising developments, existing hurdles must be taken into account. These hurdles drive uncertainty regarding the speed of development, the extent of change and the right strategies to choose in the programmatic advertising development. There are new opportunities for entrepreneurs who want to benefit from the shift.

5.1   Programmatic Buying of Premium Formats

A frequently cited problem is the availability of premium inventory and advertising formats. While large, global providers increasingly offer their formats for programmatic advertising, local vendors protect their top assets in making these formats only available in direct sales. Global providers see themselves forced to standardize global rollouts as much as possible and to offer universal formats. As a result, many of the targeted, optimized campaigns run in standardized advertising inventories, which are outside of users' awareness. Consequently, it is important to make premium inventory and premium formats available for programmatic advertising.

5.2   First Call for High Relevance

Closely related to the problem stated above is the fact that many programmatic campaigns do not have the first call in the ad server of many publishers. High relevance campaigns – usually also the one with the highest bids – need to be the highest priority in every ad server setup. It is somehow absurd that we invest so much money in finding the sweet spot in a customer's journey using all our data and technology, but cannot reach the customer with the best formats on the best inventory available. The future of programmatic is premium inventory and formats.

5.3   Required Skills and Qualified Staff

An often-cited obstacle is the lack of appropriate expertise and the right staff at media companies, marketers, agencies and advertisers. New possibilities require rethinking and technical orientation of staff previously only available at specialty agencies in Search Engine Marketing (SEM). Anyone with expertise in programmatic advertising can choose from a variety of unfilled positions. In consequence there is a high fluctuation and inflation in salaries. This drives labor costs, makes continuous software and market development more difficult and thereby inhibits overall industry development. In particular, small suppliers have enormous problems to overcome this. Industry associations and universities have to respond to the demand from the advertising industry and marketing departments of companies.

5.4 Best-Practice Strategies Beyond Remarketing

Besides a lack of know-how, there is a lack of data usage strategies to increase campaign success. What works beyond remarketing to address shopping cart abandonment and the simple selection of target group criteria for Google and Facebook?

It is necessary to overcome the experimental stage and build the necessary industry know-how on appropriate data points and data usage policies. Simple, well-established strategies are a powerful lever for the development of the industry. This includes the establishment of workflows and procedures for efficient campaign handling and the safe use of data points in programmatic environments.

5.5 User Acceptance, Data Protection and Security

Privacy and data security concerns are major hurdles that need to be overcome. Data generally offers a potential for misuse. It is important to minimize this risk efficiently and effectively. The direct marketing industry knows this too well. Despite established accepted data processing methods, direct marketers regularly suffer from data theft, spam and phishing. When in doubt, decision makers often decide to emphasize growth over security. In the online sector in general, and in programmatic advertising in particular, there is more data available than ever before. The consequence: increasing risk of misuse. An unfortunate mélange in users' perception of rather separate topics such as hacker attacks, eavesdropping government programs and data-driven banner advertising also damages the industry. Therefore, it is necessary to regain user trust with professionalism, reliability and user education. Unfortunately, a missing global legal framework, politicians focusing on making a mark for themselves and companies that act on the principle of "Shoot now – ask later," makes finding a solution extremely difficult.

6 Current and Future Drivers of Programmatic Advertising

Market trends in display advertising and the need to overcome hurdles are not the only factors that need to be considered for an estimation of programmatic advertising perspectives. There are also drivers influencing the market positively and pushing programmatic advertising to new heights and in new segments.

Most important are new emerging or transforming marketing channels, as well as new devices, previously unused technology infrastructures and newly developing formats.

7 Emerging Programmatic Advertising Channels, Formats and Approaches

Looking at programmatic advertising channels, unsurprisingly advertising in the social media environment has a high momentum.

7.1 Facebook Advertising

Based on high reach, high usage, innovative new formats and a consequent automation strategy, Facebook has succeeded in building an excellently positioned programmatic advertising environment. As a dominant supplier in social media advertising, Facebook has a similar role for social media as Google in the Search arena. Through the integration of programmatic advertising platforms and in establishing its own ecosystem with dedicated, high-performance software partners (Facebook Marketing Partners), Facebook has taken a pioneering role in the field. The Social Graph, Facebook's unique knowledge of social relations and user interests, does the final bit. The high proportion of mobile Facebook usage and efficient forms of native advertising integration give the company a strong position in mobile programmatic advertising. The entry into new formats such as video advertising and new areas like location-based advertising will provide more sustained momentum, unless the usage of Facebook diverse social media offerings will be slowed down by emerging social networks. The latter is the fundamental threat to Facebook's social media advertising ecosystem and explains the aggressive acquisition strategy of Facebook in this area. Snapchat and Instagram are now strong assets in the Facebook Family. We expect more significant acquisitions by Facebook in the coming years. With its Atlas infrastructure, Facebook will furthermore attack in Display, Mobile and Video-Advertising beyond its own platform. This will allow new possibilities to make Display more relevant.

7.2 Mobile Advertising

Due to the increasing shift of Internet usage to mobile devices, 'mobile' also plays a (if not THE) fundamental role. This is particularly of high importance as mobile use cannibalizes desktop use, forcing media and service providers to find efficient solutions. In contrast to other areas, there is an urgent need for action. Only those who are proficient in mobile programmatic advertising can remain successful in the long run. As previously mentioned, less helpful are the variety of technologies and the ever-changing definition of mobile. The currently most used definition uses a differentiation based on screen size and partially on the input interface (example: touchscreens). This makes sense; given screen size and input-interface have a relevant importance for the design and impact of advertising. However, for programmatic advertising the context of use is at least of equal importance. It is useful to differentiate out-of-home and at-home use.

7.2.1 Out-of-Home-Use of Mobile Devices

For the out-of-home mobile device use, sometimes disrespectfully referred to as "Bored-in-line" use, the local context and the interaction with the direct, local environment are most important. It is fundamental to understand a user's exact current location and with which companies or people a user can interact. Which technology will prevail – iBeacon, Wifi or even sound recognition – is still unclear, yet, for the overall development only of tactical importance. In the long run it is important to offer users programmatic value-add with a hyperlocal and relevant context. The value-add is not limited only to classic coupons, discounts and bonus programs. Also, value-added content such as user opinions or menu recommendations, intelligent assistants for decision making, entertainment offers or creative micro contents can make a mobile advertising message relevant.

If user and privacy concerns can be overcome, linking online behavioral data with contextual information from a user's real environment will be a new large mobile playing field. It is only a matter of time until for example multichannel retargeting is used massively – the short-term sales potential is too big to ignore. Especially brick-and-mortar and multi-channel merchants need to look at out-of-home programmatic advertising strategically to stay relevant with their local stores in the competition with same-day-delivery offers and comfortable packing stations in each house. It requires improved service quality in using data analysis and mobile devices to keep local shopping attractive. Even at first glance little predestined industries such as home hardware stores now have online pure-play offers competing for market shares of multichannel retailers.

7.2.2 At-Home-Use of Mobile Devices

The second major area of mobile device use is at-home, often as a second screen as one big portion of mobile device use is at home. End users often use several screens parallel. It is debatable and surely varies based on demographics whether tablets or smartphones are only a second-screen of the TV or vice versa. This is a passionate discussion between providers, mainly because of CPM prices and advertising budget allocations. However, the intelligent programmatic link in the context of different entertainment formats and online services are crucial for at-home environment campaign success. Future interaction between entertainment formats and consumers across multiple channels can be shown in today's examples such as Oreo commenting humorously on the Super Bowl blackout or Oscar selfies bringing Twitter's server to its knees.

In this playing field a decision of which value-add promotions are the correct ones and which services should be used for programmatic messages has not yet been made. It is assumed that users will prefer already used services such as

Facebook, Google and Amazon to specialized providers. Already, communication on Facebook and WhatsApp today wins over communication in special second-screen applications from start-ups and television stations. As a good working hypothesis it can be noted that in low-involvement programs, users do other things and advertisers have to raise users' program awareness with intelligent programmatic use. In high involvement entertainment such as talent shows, sports broadcasting and computer games, it is important to intelligently link in the program with added value.

At-home-use does not have to be limited to screens. Apple, Amazon, Microsoft and Samsung have been experimenting with voice recognition at home for quite some time. Although voice recognition for phones, TV, sound devices and game consoles are still of limited value due to a lack of recognition quality, we are getting closer to voice recognition that works. Voice searches will become an interesting field for programmatic advertising messages; voice orders via intelligent audio or a smart watch can become a powerful sales channel.

7.3 Online-Video and Digital Out of Home Advertising

The fact that young audiences consume video content increasingly less as linear television but rather in on-demand online video channels and social networks opens other large spaces. The emergence of new video content formats and YouTube stars puts pressure on companies. For example, a trendy brand wants to be associated with the stars of the digital natives. Brands increasingly rely on online videos as a complement to TV and Display Advertising, or sometimes even as an alternative.

In addition to reaching new or lost target groups, online videos promise a better advertising effect than traditional display formats. With an increasing reach this field will be interesting and highly relevant for programmatic advertising strategies. Advertisers focusing on the out-of-home environment have realized the importance of video, too. Using digital advertising space ("Digital Out of Home", (DooH)) in railway stations, airports and shopping centers, advertisers focus on video content and commercials beyond the TV screen. With the build up of ad-servers and in combination with mobile users' location, a new programmatic playing field is defined "further down the road" – the combination of mobile devices and DOH. A small but steadily growing number of TV objectors will have to be brought in contact with highly targeted video advertising through other channels.

7.4 Content Marketing and Native Advertising

The developing field of content marketing and the closely related field of native advertising offers programmatic potential. The creation of own 'owned' media formats is still high on the agenda of many companies, but lines are blurring. Data driven media companies such as BuzzFeed already show the future of a data-driven and individual user focused content distribution. Programmatic marketing gets a new meaning in a content marketing context. Programmatic brand responses on latest news such as the mentioned Super Bowl comments are a special form of programmatic marketing and are increasingly cultivated by brands.

Data-driven content marketing in programmatic not only can revolutionize dissemination; there is also potential in content sourcing. Started by companies like Demand Media with low cost, mass production of content for specific keywords, this can continue in the programmatic area. Why not build large content databases, which disseminate content via programmatic systems to potentially interested users and reward authors only based on clicks? Why differentiate still between content of media company editors and brand editors? For example, for content marketing pioneer Red Bull, boundaries between Media Company and brand are blurred. Platforms such as LinkedIn and Facebook have become top traffic content suppliers with their dynamic content recommendations in mails or their news stream. As often during the Internet's development intermediaries will gain more power and become more important traffic sources.

8 New Data and Findings as Programmatic Advertising Drivers

The last example leads us to another driver of the programmatic advertising development: Emerging or available user data and the resulting findings generated by companies.

8.1 Big Data Analysis

Factors driving this trend in recent years are the often quoted Big Data scenarios and related Big Data initiatives of companies. The explosive increase in data through digital media, mobile devices and social networks is facing a huge increase in data processing capacities. New database technologies allow for a more cost-effective and efficient information processing and thus enable new customer insights. Beyond these insights programmatic, data-driven advertising workflows lead to an unprecedented speed in analytics and individuality of advertising messages.

While most companies are still struggling with the sheer enormity of processing the data, other companies already use relevant findings drawn from the environment. Machine learning thereby will enable systems to derive better and better forecasts. Predictive approaches today offer a quality level, which is still too low; however, laboratory developments promise great progress. In experiments lies were detected with frightening accuracy through facial expressions. Leading thinkers such as Bill Gates have recently pointed out expected negative consequences of machine learning on the job market in the future. Intelligent algorithms will replace many jobs currently done by well-educated knowledge workers. Providers in the programmatic advertising environment without a doubt will use such available technologies to gain a competitive advantage.

8.2 Web of Things and Virtual Reality

With the proliferation of the Internet of Things, ("Web of things," the connection of an almost unlimited amount of networking devices on the Internet), this trend will accelerate again. Experts estimate the number of Internet-connected devices in the future will be several fold the number of currently connected devices. The volume of machine-to-machine communication will by far exceed that of human communication. Spectacular acquisitions such as the purchase of NEST, a home automation manufacturer, through Google or the launch of Google Ara Project, an open free smartphone platform, spur imagination and emphasize the importance of the web of things.

A simple look at the Apple Store already shows which enormous data potential already can be used. No matter if sleep patterns need to be improved, athletic performances recorded or blood pressures measured, using affordable, easily available accessory components, everything can be captured that is economically measurable. If all user appliances, the car, house, smartphone etc. provide valuable data points, new opportunities for programmatic advertising will open up. This is not a look in the distant future. Already today data of almost all cars driving on the streets can be accessed via a simple standardized on-board computer interface and continuously evaluated via smartphone. There is no need for complex installation; a kit of Automatic for under $ 100 for self-installation is adequate and provides an amazing variety of data and application cases. Are you looking for your car? Here it is. You drive more than three hours on the freeway? Have a break and get a KitKat with a 50 % discount code. Expectations need to be recognized, then offers can be directly provided.

9 Risks in the Programmatic Advertising Development

Despite all excitement about programmatic advertising possibilities, risks for programmatic advertising development must not be ignored.

9.1 Restrictive Regulation of Data Protection

First, there is the risk of a too restrictive or even prohibitive regulation of data use. It is popular among politicians to protect consumers from a supposed, unfortunately sometimes very real, industry "exploitation." More and more citizens perceive the use of their data as being close to the edge or even illegal. The added value of a more relevant communication and customized services is not recognized, or the price – giving up parts of privacy – is considered too high. The legitimate demand for a regulation of new emerging data privacy issues is used by some populist politicians to call for a ban of online advertisers "spying on users." The industry being hesitant in self-regulation did not win laurels in the past. It is therefore a legitimate fear that restrictive regulation could have serious consequences for the further development of programmatic advertising. The IAB and major industry players need to address these issues with a higher priority and be less focused just on their interests. We need a social contract for data usage in order to keep programmatic advertising thriving.

9.2 AdBlocker as Self-defense of the User

In their weariness about annoying ad formats and effective but fairly aggressive programmatic advertising approaches such as remarketing, users increasingly take the initiative. They install ad-blockers and regularly delete their cookies – those text files that are fundamental for targeting technologies. Industry is responding inconsistently, ranging from resignation and frustration to education campaigns and paid-service initiatives. Behind the scenes there is a technical arms race happening, examples are anti-Ad-blocker services or new identification technologies such as fingerprinting, the cookie-free detection of users according to specific characteristics such as their PC configuration. There is no evidence for the success of these technical approaches yet. They are also likely to be questionable in the future. Although it can be assumed that ad-blockers will not be the undertakers of the digital advertising industry, their use can mitigate overall digital campaign performance (especially programmatic advertising) significantly. Again, best positioned are global providers using own login services compared to companies depending on cookies and third-party providers.

9.3 Limited Branding Capabilities

In order to bring Digital Marketing and Programmatic Advertising to the next level of development, we need to solve the branding problem. Despite many efforts Digital Marketing often fails to deliver results for branding. On the other hand, this can be attributed to the way we use digital media. Most of the time we use it as a service, searching things, skipping through news streams, buying stuff, doing communication – getting things done. Not a good branding environment. Whenever we are in entertainment mode we switch to On-Demand Video and Games. A far better branding environment, but often monetized through item sales and paid content instead of advertising. Beyond YouTube and increasingly Facebook we find thin air. On the other hand, the branding problem is caused by the digital industry itself. We manage what we measure. We track clicks, optimize conversions and decide based on order lifetime values. These are Sales KPIs and we are therefore seen as a sales channel. We fail to measure and optimize on Branding KPIs, we are even afraid of these KPIs. Brand Marketers want transparency on viewabilty and we get nervous and are slow to offer a solution. To bring Programmatic Advertising to the next level we need to embrace branding KPIs; we need to measure Engagement and allow optimization toward brand perception. Data-Driven Marketing can do a lot for brands; brand enthusiasts could be addressed with different content or marketing messages than average users. Programmatic approaches could be used to add value and interactions with the brand. Instead of the meaningless "thank you for flying Lufthansa" after landing, the airline can thank loyal customers personally on their mobiles and give them added value through deals or content. If the flight is late, we could give an honest apology and try to make it up by upgrading their Uber ride. Comfort features like direct flights could be advertised to travelers waiting in line at transfer terminals. These customers will remember the high amount of direct connections offered by Lufthansa far better.

10   Conclusion and Outlook

In summary, programmatic advertising is one of the most dynamic areas within the megatrend marketing automation and the prospects are very good. The increasingly established ecosystem for programmatic advertising, strong drivers such as Big Data and emerging new marketing channels face relevant risks, which need to be addressed, and hurdles, which can be overcome. These hurdles and risks are not impossible to overcome. It can be assumed that the discussed privacy problem will be relevant for years, but in offering real added value for users and an effective self-regulation this problem can be contained. Who would have thought years ago that today a majority of smartphone users willingly and continuously allow themselves to be located, and that they would be okay with transmitting this highly sensitive information using apps?

The competition among market participants and marketing channels will be exciting in the next few years. Due to the high complexity of the new ecosystem, still quite low entry barriers, and low transparency in the market, there are many new market players. An opaque ecosystem of niche providers and platforms has developed. A consolidation similar to the Google-Search-ecosystem can be expected. Major platforms increasingly offer own functions and so many niche vendors will be obsolete or are pushed into a new role, being a supplier of functions in an App system. As market participants, the global providers Google, Facebook and Amazon are in a good starting position. Global agencies also have great opportunities because of their expansive abilities and power; however, the amount of existing assets (for example their reach) is limited. It remains to be seen how much specialized providers, which again have access to substantial capital resources, will enter the league of top players.

Regarding marketing channels, it is important to quickly look beyond the display-advertising horizon. Although programmatic advertising develops very well in the area of standard advertising space, new marketing channels such as social media advertising also show an impressive dynamic. Video advertising and native advertising channels are strongly sought after formats, particularly because of banner blindness (users ignoring banners). A mobile strategy is a must, not an option. The user and Google already punish you if you don't have a mobile site, other channels will follow. It is expected that programmatic systems advertising can and must use a variety of different channels. One of the most fundamental changes in the implementation of programmatic advertising has to be the shift from a campaign orientation to marketing programs. What is common and useful in the search area, the setup of continuously running marketing programs, is for all automated programmatic channels a critical success factor. Of course, campaigns will continue to exist. There is always a manager who finds an opportunity or needs a push to achieve certain company goals in a specific timeframe. More and more marketing budgets will be used in successfully tested programs for the differentiated approach of old and new customers in appropriate channels. We also need to find best practices in creating value for brand markets using programmatic approaches. Far more efforts and resources need to be allocated to this problem. Clicks and conversions are easy to measure, but focusing on these KPIs and ignoring brand metrics and thereby also brand budgets limits the Programmatic Industry growth.

A functioning programmatic advertising ecosystem will have to efficiently use all relevant channels at the end of the day, to effectively address the still unresolved problem of cross-channel and cross-device adaptability. Through programmatic buying more offers and channels can be efficiently addressed in the future. At the same time, the assessment of individual channels is increasingly adapting to the real advertising result. Overfunded media will face reductions. This can reach the point of profitability and requires radical rethinking and courageous actions.

Despite the overall added value of diversity, a consolidation in the programmatic industry is a good and necessary development. Numerous specialist providers in the ecosystem solve many customer problems. However, various "feature companies," companies whose business model is based on individual software features, rather slow down market development in the medium term. Furthermore, too many small businesses involved in sales lead to distortions in the value chain. At the end industry should not forget that everything we do is to connect businesses with customers or potential customers. There should not be too many market participants in between. Hopefully, through consolidation, one or the other Non-US player succeeds and still there will be enough variety and innovation.

Prof. Dr. Jürgen Seitz is Professor of Media, Marketing and the Digital Economy at the University of Media in Stuttgart. The research focus of Professor is digital marketing, online business models and the early phase of innovation. In his professional career he worked among others at Microsoft, WEB.DE and in the United Internet Group, most recently as Managing Director of United Internet Dialog GmbH.

Steffen Zorn is the Program Director Marketing & Management at Swinburne Online, a partnership between Swinburne University of Technology and SEEK, in Melbourne. He holds a PhD from the University of Western Australia and a Master of Business Administration. He has taught digital marketing and marketing management related units, and has researched online consumer behaviour and customer lifetime value topics. His work experience includes positions and projects in automotive, machine building and related service industries.

Part II
Components

Consumer-Centric Programmatic Advertising

Oliver Gertz and Deirdre McGlashan

Programmatic Advertising has evolved from simple Retargeting and Audience Buying into consumer centric marketing against the entire purchase funnel and the method of choice to execute all digital media buys for many marketers.

This increased use and sophistication has led to discussions about the business models of Programmatic Advertising solution providers that range from packaged products to customised fully disclosed services.

Each programmatic setup combines a DSP, First and Third Party Data, Media Inventory, Brand-Safety approaches, Adserver and last but not least experts that drive the whole system. Advertisers have more and more choice on configuring their own setup which moves the discussion away from "agency vs. in-house" to "what is the right combination of in-house and outsourced for our needs."

1 Programmatic Advertising in the Marketing Mix: Content and Connections

Present day marketing requires that marketers consider the entire system of communications and how to optimise this to step change their business outcomes. In other words, we live in the age of Content and Connections – understanding how channels and content can work best together to increase the productivity of communications. Content is the fuel for high-performing systems. This means any form of consumer messaging that connects brands to their consumers, whether that is TV, press, radio, events, partnerships, branded content, social, search, mobile

O. Gertz (✉)
MediaCom, Medienbrücke, Rosenheimer Str. 145d, 81671 Munich, Germany
e-mail: oliver.gertz@mediacom.com

D. McGlashan
MediaCom, 124 Theobalds Rd, London WC1X 8RX, UK
e-mail: deirdre.mcglashan@mediacom.com

or new technologies. Connections is how we connect the content to the consumers around the system. It is about having a clearly defined distribution strategy, ensuring there are no dead ends or wastage resulting from disconnected content.

Programmatic Advertising is a huge leap forward in our ability to deliver the right content to every single consumer to drive the most impactful relationship. The combination of real-time technology with a more and more holistic understanding of the individual user-profile allows to deliver on the promise of One-to-One communication.

1.1 The Evolution of Programmatic Advertising from a Performance Channel Only to a Method to Buy a Broad Range of Digital Media

The first incarnation of Programmatic Display Advertising was adopted under the banner of Real Time Bidding by performance driven advertisers.

Programmatic Advertising has come a long way since the early days of performance-driven Real Time Bidding. We can look at it as the evolution of Programmatic Advertising from 1.0 to 3.0 (Fig. 1).

Re-targeting is a very powerful tool to re-engage with consumers that have shown an interest but have not converted. Targeting data profiles are collected through a very simple procedure of placing pixels on the website along the path to conversion, usually only looking a very few data segments (however often providing thousands of customised offers based on the last product the consumer looked at). The main requirement for the media buy is near unlimited reach since finding those few thousand consumers again amongst hundreds of millions of internet users requires seeing billions of ad-impressions across thousands of websites.

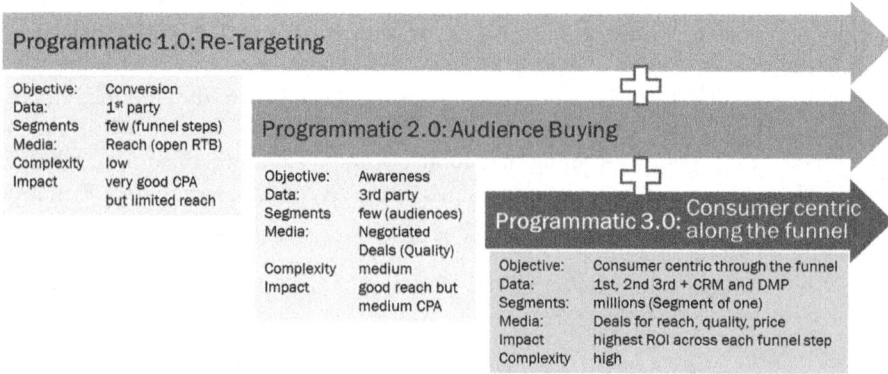

Fig. 1 Programmatic advertising 1.0: retargeting (Own image)

Real-Time-Bidding on open exchanges is a very efficient way to buy that reach, and the quality of the placement is secondary to the chance of finding the right user profiles again.

Re-targeting produces a very good Cost per Action but has limited reach in the media mix, as it applies CRM-approaches to re-engage with the small segment of potential customers who have shown an interest to purchase.

1.1.1 Programmatic Advertising 2.0: Audience Buying

While Retargeting focusses at the very end of the purchase funnel, Audience Buying is more applicable for the upper funnel to reach consumer profiles that have not engaged with the advertiser yet.

Programmatic Products such as Quantcast, XAXIS or Rocketfuel collect a vast amount of anonymous consumer data to build targeting profiles. When Audience Buying, the media bought is audience focussed, trying to reach the right consumer segments, rather than buying ads in particular editorial context.

When Audience Buying, advertisers are much more concerned with the media placement quality than with Re-targeting as the media output – reach in a defined audience segment – is a proxy for the campaign goal of driving awareness, purchase intent and (indirect) purchases.

1.1.2 Programmatic Advertising 3.0: Consumer Centric Advertising Along the Entire Purchase Funnel

The great Return On Investment gained from Retargeting and Audience Buying has convinced many advertisers on the value that Programmatic Advertising can play against the entire purchase funnel, and the value that both data and placement quality add.

In Programmatic Advertising 3.0, advertisers move to a Consumer Centric view. They combine as much data as possible into a holistic user profile, combining sophisticated segmentation of the website visitors with CRM and even sales data and enriching it with 2nd and 3rd party data.

Those profiles are often applied to a much more selective media buying process, looking at inventory quality (brand-safety, viewability) and a more qualitative publisher selection. This can be facilitated through Private Marketplace Deals (PMD) that combine the benefits of direct negotiation (more control of where the campaign runs, access to quality inventory and premium ad-formats and negotiated pricing) with the benefits of data-driven cherry-picking of impressions in real time.

1.2 Consumer Centric Marketing: The Single View on a Consumer

Consumer-Centric Programmatic Advertising is quite a radical change in the approach to media planning. Traditionally, media planners allocate budgets across media channels (on- and offline) to selected placements (based on affinity, reach or historical performance) to buy impressions that will hopefully reach the right consumer segments (Fig. 2, *left*)

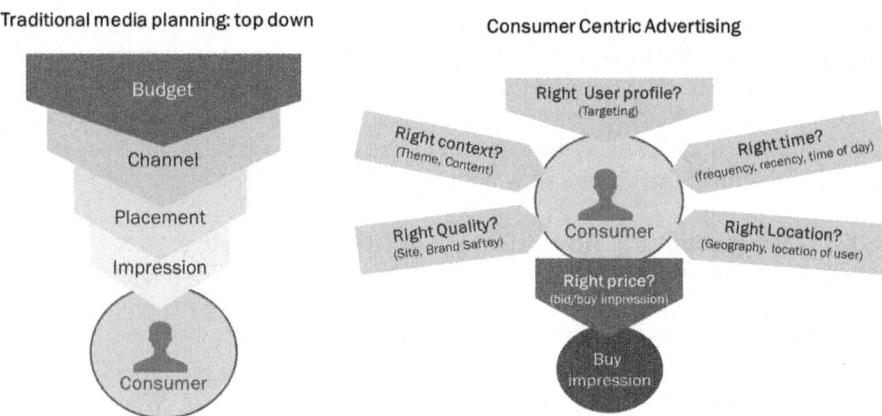

Fig. 2 Consumer centric advertising versus traditional media planning approach (Own image)

The New Consumer Centric Advertising Model (Fig. 2, *right*) Starts with the Consumer.

1. Profiles are built from all available data sources, combining first, second and third party data
2. When the DSP (more detail on Demand Side Platforms in Sect. 3.1) sees a consumer profile, complex decision processes evaluate where that single profile stands on the path to purchase, what the value of that profile is and therefore what the value of an impression would be right now. For that calculation, the consumer profile data is overlaid with additional data points to answer the following questions:
    (a) **Right context?** What is the content of the page, based on the publisher content classification, semantic content classification tools or hand-picked pages
    (b) **Right quality?** Based on blacklists, ad-verification tools or hand-picked whitelists of sites
    (c) **Right time?** Analysing the time of day, day of week but also how many contacts that consumer profile has been exposed to date (frequency) and when the last contact occurred (recency)
    (d) **Right location?** With more and more contacts happening on mobile devices, the exact location of the user may have huge impact on the receptivity for a specific piece of content. Simpler forms of location targeting may be city, postcode or simply avoiding users that are out of the country
    (e) **Right price**? All the data-points above combined with an ROI forecasting model determine the value of delivering an ad to this profile on that site right now, which informs the bid price that ultimately leads to buying the impression or not.

This new Consumer Centric model often requires a significant re-organisation of the digital data collection process to combine data that may sit in siloes right now. It is fuelled by attribution models that can assign value to each touch-point and also requires a planned organisation of campaigns.

Rather than running short-term campaigns for specific products or offers, Consumer Centric advertising is always-on, the product and offers are selected based on known consumer interests rather than marketing top down decisions.

As in Search Engine Advertising, this leads to a different process with constant refinement of the core campaign, while more traditional bursts of push communications are added to the always-on campaign for agenda setting and storytelling.

2 Configuring the Right Programmatic Setup for an Advertiser

The Programmatic Advertising industry is challenged by the debate about transparency and control, openness and walled gardens, and the value that the many players add to the ecosystem.

Understanding the marketplace and options, it's clear that there is no single "right" or "wrong" approach but only a "good" or "not so good fit" to advertiser's needs. Getting clarity on the options and how different and sometimes conflicting objectives can be balanced against each other is key. The objective of controlling inventory quality very tightly, for example, conflicts with the objective of gaining the highest reach at the lowest cost.

Understanding the different business models of the many programmatic partners on the market and which components build up a programmatic setup for an advertiser will help to make the decisions most appropriate for a specific advertiser's need.

2.1 Programmatic Business Models: Products, Services and In-between

Many programmatic buying options presented to advertisers today are an evolution of the ad-network model and can be described as a **programmatic product**: the advertiser or their agency buy a defined product such as clicks, leads or audience reach that is delivered at an agreed price by combining media inventory, data, technology and a service layer. These products are "undisclosed," meaning that the seller does not disclose the individual prices of each component but only a final price, similar to how media has traditionally been sold. These products are undisclosed but can be transparent in that the advertiser completely understands what is purchased, what it includes and price benchmarking and rigorous ROI measurement vs other products in the market is done.

The benefit for advertisers is the guarantee of the agreed output (i.e. clicks, leads or audience reach). In addition, using a single programmatic product offers simplicity of the media buy and the vendor often can offer cost benefits through bundling of media, building user segments that are made available to all advertisers and by simplifying the internal processes through productisation.

At the other end of the spectrum is the customised **programmatic service model**, where the provider (usually an agency or technology vendor) sells a service that utilises technology and data while all components are charged separately in a fully disclosed cost model. The advertiser has full control and transparency, but usually also carries the risk of buying impressions on a CPM that then need to be converted into the desired outcomes (clicks, leads, sales, audience reach). When using multiple programmatic products and/or as more and more media becomes available programmatically, the programmatic service allows for optimisation across different products.

And in the middle, there are **hybrid models** that may look like a fully disclosed service model but have mark-ups or arbitrage built in to make the service appear to be at a very low-cost.

2.2 Transparency and Disclosure

In the public debate, transparency and disclosure are often used as synonyms, however they describe two quite different concepts.

Transparency means the Advertiser understands what they are buying (format, placement, and quality), what the value is (ROI or benchmarking to similar product on the market) and how it was "produced" (i.e. inventory sources, data sources, technologies used, intermediaries).

Disclosure means the programmatic provider discloses the cost of the inventory, the 3rd party technology and data and the cost of the staff as an agreed rate.

Programmatic products tend to be transparent but not disclosed and programmatic services can be either disclosed or not. In both cases, it is important to have a strategy, benchmarking, ad-verification and robust ROI measurement. In the transparent model particularly, knowing how much each partner earns may be interesting, however evaluating how much value each partner adds to the desired results is much more important.

2.3 Advertiser Control Versus Guaranteed Results

Advertiser demand for simple solutions and low risk have led to the growth of programmatic product offerings, especially in retargeting, performance marketing and audience buying.

The advertiser agrees a defined outcome with the vendor and the vendor buys media inventory (usually at fluctuating auction prices), adds technology, data and service and sells at an agreed CPM, CPC or CPA. The vendor takes the risk of media converting to the agreed outcome but also keeps the optimisation gain, which is sometimes called arbitrage.

Advertisers who want to take full control of each component's cost also must take on the full conversion risk, buying at auction CPMs and converting into the desired outcome but then also capturing the entire optimisation gain.

It is impossible to have both: simple risk-free products and full control of the revenue each partner in the value-chain makes.

As the Programmatic share of media spend grows, first-party driven consumer centric tactics are applied along the whole purchase funnel and advertisers develop a deeper understanding of the different data and technology levers, the desire to take more control tends to also grow, and a combination of control and simple risk-free products tends to be used.

3 Configuring an Advertiser Programmatic Setup

Regardless of the model (in-house, agency, product provider) that an advertiser choses, the programmatic setup is composed of the same components of technology, data, media inventory and experts that are combined to deliver a programmatic campaign.

As the programmatic landscape and ad-tech is evolving quickly, advertisers have a lot of choice to source these components and combine them into a customised setup.

3.1 DSP: The Buying Engine

The Demand Side Platform is a core technology for any programmatic advertiser as it is the tool though which programmatic inventory is bought from exchanges and SSPs (Supply Side Platform) based on the user profiles the advertiser has acquired.

There are many players on the market that offer DSP solutions and the continued inflow of venture capital means that the landscape is evolving quickly.

DSP selection criteria	Comment
Access to all inventory sources	The DSP should be connected to all relevant SSPs and exchanges in the market, which sometimes can be local players
Access to event-level data	To maximise the value of the volume of data generated, advertisers may want to extract event-level data for each impression into their own analytics tools, such as for attribution modelling
Integration of 3rd party tools	Flexibility to bring in specialist tools to adapt for new opportunities. Example tools are ad-verification, dynamic creative optimisation, data tools, etc.
Control over algorithm	Algorithmic optimisation plays an important part in Programmatic Advertising and the advertiser should have as much control as possible over how the algorithm optimises

(continued)

DSP selection criteria	Comment
Awareness of vendor owning media inventory	A technology vendor that is also selling own media on the platform could be conflicted to either provide independent buying across all media vendors or increasing the share of the vendor's own media. In this case, it is important to have a clear strategy, benchmarking and robust ROI measurement
Vendor earns only licence fee	If control is key, the vendor should only earn money on the licence fee paid by the advertiser and not by taking a position in the media itself

3.2 First Party Data: Maximising the Value and Control of Owned Data

First party data is the data that advertiser generates through direct consumer interaction on his owned assets, CRM and customer databases. It is the most valuable data as those consumer profiles have already engaged with the advertiser so are therefore usually much more likely to convert into the desired action. The advertiser knows which offers they have looked at and where they are in the purchase process. Additionally, first party data is free beyond the cost of collecting it.

The greater the amount of first party data collected, the more concerned advertisers are about data security, ensuring that the data collection and usage is compliant to privacy laws and industry self-regulation, but also ensuring that the data is not used for other advertisers.

First party data requirements	Comment
Advertiser is the owner of the data	In most programmatic configurations, some advertiser's data will physically sit on third party platforms (DSP, DMP, Adserver, Retargeting Ad-Network etc.), so contractual clarity about data ownership is essential
Data share with other advertisers	Data clauses should be very clear how the vendor can and cannot use the advertiser's data so that their data is not shared with other advertisers without their knowledge or permission
Privacy is ensured	Data collection, profiling and usage is governed by local privacy laws. Additionally, industry self-regulation for Online Behavioural Advertising should be adhered to
Integration with a client Data Management platform	Advertisers may want to build an own Audience Data Management Platform that all consumer data is integrated into

3.3 Third Party Data: Worth the Money?

Third party data is data that can be rented from data vendors. Data marketplaces and most DSPs provide connections into many data vendors that usually charge a CPM-based fee for the use of the data.

There are other data sources beyond audience data that often work very well: contextual data to evaluate the content of URLs in the bid-request, location data for the user, external data signals such as weather or signals when a TV or radio spot is on-air can fuel much better media buying decisions.

The availability and quality of third party data varies by country and target audience. As data vendors often combine data from any different sources and collection mechanisms (that are sometimes not fully disclosed), it is often hard to predict how well those data segments will help to increase the targeting quality, measured by an increase in achievement against objective.

Third party data requirements	Comments
Documentation of collection process	Data vendors should be as specific as possible to explain how data was collected and profiles are constructed
Privacy compliance	Data must be collected according to local privacy laws but also follow industry self-regulation best practice
Data quality	Disclosure of data collection process, freshness of data and the methods of segment building help to evaluate the data quality
Test of data ROI	Structured test of different data sources against same media help to evaluate the ROI uplift of Third Party Data segments and whether the cost are justified

3.4 Access to Media Inventory: Open Auctions or Private Deals?

As Programmatic buying moves beyond Performance marketing, advertisers put much more emphasis on the "quality of inventory". They want access to premium inventory, high impact ad formats and to be rewarded with lower prices for spending higher budgets or even going into spend-commitments with publishers. And they want to know where their ads run and work with publishers they know.

Publishers, on the other hand, also want to know the advertisers that appear on their site, they want to keep control of their pricing strategy (direct sales vs. indirect, guarantee vs. auction) and they prefer to know how much they will earn more than 120 milliseconds before the ad space is sold.

Programmatic Advertising 1.0 (see page 1) was all about low-cost reach to generate a high volume of low-cost conversion for advertisers. Publishers saw it as a method to sell remnant inventory. Both saw Programmatic Advertising as a new niche channel.

Programmatic Advertising today is seen as a method to trade media, growing into the method of choice to trade all ad-served media. Therefore, all needs of media selling and buying have to be catered for, from high-end premium to low-cost reach.

This can be achieved through Private Marketplace Deals. They combine the benefits of traditional media trading (mainly long-term relationships between buyer and seller, individual discussion about which placements and formats to buy and negotiated prices) with the benefits of Programmatic Advertising to buy the individual ad impression in real time based on data.

Programmatic trading deals are the preferred route for large brand advertisers and their agencies and access to the valuable data that many publishers have about their users is more and more becoming a core focus area.

Media inventory requirements	Comments
Access to all inventory sources, display, video and mobile	Advertisers want access to all available inventory, all publishers and all media types. They need DSPs that are connected to all SSPs and Exchanges in the market, cover all screens and support Private Marketplace Deals
Buy from the source – cut out the middle men	Advertisers want control and savings – buying as closely as possible from the source (publisher or their chosen SSP) is preferred
Private marketplace deals	PMDs are a core component to combine the benefit of negotiated packages with the benefit of real-time execution
Relationships and trust	Advertisers want to know whom they buy from. They want to work with respected partners and be rewarded for higher spend levels
Open exchanges and reach	Open exchanges are not irrelevant, they give access to a vast amount of long-tail inventory and low-cost reach if ad-verification is applied prudently

3.5 Brand Safety: Preventing Fraud, Controlling Viewability and Audience Reach

Private Marketplace Deals and close control on what inventory an advertisers buys from whom will also address one of the biggest challenges of Programmatic Advertising – brand safety.

Shocking reports on fraud, low viewability and brand safety claim that 57.4 % of ads bought via networks and exchanges are not viewable, that 14.5 % of those impressions are fraudulent and 17.5 % have a brand safety risk.[1]

---

[1] Source: http://integralads.com/wp-content/uploads/2015/02/Integral-Ad-Science_Q4-2014_Report.pdf.

Advertisers and their agencies have many tools to address brand safety. Applying blacklist of known "bad" sites is the base level that should be done but has limited long-term effectiveness as fraudsters launch new sites every day. Therefore these lists must be updated frequently. Ad-Verification tools that analyse the page's content and block a bid on questionable sites are an effective addition. A whitelist of hand-picked sites and partners gives the highest level of control but limits reach and usually increases eCPM.

Beyond brand-safety and fraud, verification tools can measure ad viewability and audience reach vs. a panel.

Controlling brand-safety	Comments
Blacklist	Block known "bad" sites – necessary but not sufficient
Ad-Verification tools	Detect fraud and pages that are not brand-safe, allowing advertisers to define their own rules for "what is safe"
Whitelist	Hand-picked list of sites to advertise on (and from whom to buy them). High control, lower reach and usually higher CPMs
Viewability tracking	Low viewability must not be seen as fraud unless the seller has guaranteed viewability, but optimising for viewability is very important for brand impact and conversion
Audience verification	Measures demographic reach against a panel. Panel-size may be a problem and demographics are often not the most relevant description of a target audience, so use with consideration

The right combination of those tools gives advertisers a level of control and safety that often goes beyond what is established today in traditional display buying. Some of the most shocking examples of brand-damaging ads happen on respected news sites, where airline ads appear next to plane-crash reports or confectionary ads next to articles about obesity. Advertisers who apply pre-bid blocking verification tools can define their own keyword cluster and semantic rules to enforcer their own definition of brand-safe. And some verification tools now also include the analysis of pictures and videos to detect context and dangerous content.

3.6 Customising Messages: Dynamic Creative Optimisation

Consumer centric marketing with deep segmentation and laser targeting is not effective if the message is not customised to the consumer and their needs.

Customised content for consumers against their needs can be done through multi-variant creative, (creating many different versions against the anticipated audiences and need states). More powerfully, Dynamic Creative Optimisation (DCO) allows the automated assembly of ads from multiple components (copy, image, call-to-action, offer etc.) to give each consumer the right message at the right time to drive them to the most effective reaction.

The same data that informs the decision for whether or not an ad impression should be bought on a specific site for a specific user can also inform the decision for which creative and message to show to that user.

Simply showing the same products that the user has looked at a high frequency may generate some clicks and sales but also alienate many consumers as they may feel stalked. So finding the right messaging strategy that can support brand story-telling and gently pushes the user down the path to purchase will be the task ahead and lead to a new level of collaboration between creative minds and math men.

3.7 People: Evolving Skillsets for Granular Optimisation of the System

In the early days of Programmatic Advertising, the public discussion was very focussed on technology and automation. Some still believe that they only need a login to a DSP, push some buttons and then the algorithms do the rest.

While some reach-focussed tactics like retargeting and audience buying benefit from large data and machine-learning based automated optimisation, the need for smart people has not decreased but is in fact increasing. The fast evolving landscape, the application of Programmatic Advertising along the whole purchase funnel, complex audience segmentation, Private Marketplace Deals and dynamic creative optimisation, constant management of brand-safety and optimization of the whole system toward maximum ROI is a complex task.

Advertisers and agencies have learned that the workload of complex programmatic campaigns is higher than traditional display buying as so much more data can and needs to be analysed, used and optimised.

In traditional display buying, the team sees the number ad-impressions, clicks and conversion per media buy and can change the creative or cancel the buy. In Programmatic Advertising (within the service, hybrid or product models), the team can get full transparency about every site, every impression, every user segment, optimise in real-time based on those, time-of day, day of week, geography and more. Algorithms will only pick up after long learning periods and human optimization still beats the algorithm in many cases.

It is similar to Search: in theory a simple to manage media buy with often a single vendor, very few ad formats, limited targeting capabilities, brand-safe placements and established ecosystem of tools and experts that have evolved over the last 15 years. However, everyone agrees that Search is more labour intensive than traditional display buying as it generates much more granular data that can and needs to be optimised. While Search may be twice as labour intensive as traditional display buying, Programmatic Advertising is at least three times more complex.

So we see the evolution of digital media planners into programmatic planners and the addition of data consultants and data scientists into the teams.

Consumer-Centric Programmatic Advertising   67

Programmatic experts requirements	Comments
Programmatic consultants	The evolution of the digital media consultant helps to understand the landscape, opportunity and risks and define the Programmatic strategy
Programmatic planner	Builds the plan, defines KPI and benchmarks, inventory strategy, audience segmentation, messaging plan and optimisation rules
Programmatic optimiser	Is the hero in the team, working on the granular levers to drive Return on Investment by optimising based on deep analytics of all available data
Inventory manager	Negotiates private marketplace deals, monitors brand-safety and inventory quality and works with publishers on innovative formats and data segments
Data consultant	Works with the web-team, CRM, DMPs, software engineers and data providers on the "data plumbing" to ensure seamless data flow while protecting data ownership and privacy
Data scientists	Analyses the data with statistical models and machine-learning algorithms to generate deeper insights and optimisation rules

4 Moving into the Future

As the Programmatic Advertising landscape is evolving very quickly, it is difficult to determine which trends will have the highest impact on how advertisers buy media.

We believe three topics will dominate the discussion for the near future:

1. How does Consumer Centric marketing change the structure of how advertisers and agencies are organised?
2. The discussion within advertisers about taking control of Programmatic Advertising by taking it in-house.
3. How Programmatic Advertising will change the ways TV is planned and bought.

4.1 The Consumer Centric Organisation Does Not Separate Marketing and Sales

In a Consumer Centric marketing organisation, the consumer determines which offer he or she gets when and where. The advertiser looks at the consumer profile, analyses the context and time that the opportunity to show an ad is offered and tries to forecast which offer and message is the most relevant and what the value of delivering an ad to that consumer profile is. All those data points determine whether an ad is shown and which message is displayed.

An organisation that is siloed into marketing and sales, delivering campaigns based on what a marketing or sales manager wants to communicate and then broadcasts that message as a top-down push message is not set up to capitalise on the Consumer Centric opportunity.

The advertising organisation of the future is built on a unified view of the consumer, bringing data from all touch points together and building consumer segments based on their needs and value. Then each segment gets the message and offers most relevant for them – down to the segment of one. Communication is always-on, based on consumer needs and receptivity, but also enhanced with agenda-setting push campaigns.

Data fuels the media buying decision and message delivery. That does not mean, however that big ideas and creativity are not relevant anymore. Programmatic technology allows a much more granular campaign delivery, but understanding consumer needs and motivations and turning that into a communication strategy based on a big idea and engaging content will be as important as ever.

Once that big campaign idea, visuals and story have been developed they then can be adapted for all consumer segments and formats in dynamic creative formats. Programmatic Advertising can and should support more personalised storytelling.

Measurement of results is more important than ever, attribution modelling helps advertisers to understand the contribution of each digital contact to short term results while econometrics track the cross-channel effects and long-term impact.

4.2 Should Advertisers Take Programmatic Advertising In-house?

A debate is going on in blogs, in conferences and within the industry whether or not advertisers should take Programmatic buying in-house. This discussion is driven by increased advertiser sophistication and sometimes motivated by frustration of non-disclosed product models and the lack of transparency in the whole ecosystem.[2]

Now disclosed service models are available in the market that combine the benefits of full advertiser control with the benefits of outsourcing (some or all of) the complex operations by tapping into large specialised programmatic teams and/or incorporating programmatic products.

The technology stack needed (DSP, verification tools, an Audience DMP, etc.) are now available that give advertisers and agencies a lot of choice and flexibility in a highly competitive and fast evolving tech landscape to select the best choices for the objective at hand. 'The real differentiator of any programmatic setup are therefore the people that operate those systems, their expertise and resources, relationships to media and data owners and their ability to drive results.

As agencies like MediaCom now offer customised Programmatic Buying Unit services that can plug into programmatic products to advertisers, the difference of

---

[2] One Example is the WFA Guide to Programmatic: http://www.wfanet.org/media/programmatic.pdf.

outsourced vs. in-sourced turns mainly into a people and talent discussion – like other outsourcing discussions before.

	Advertiser in-house	Outsourced
Technology stack	Full control and choice Mainly (combination of) off the shelf products	Can collaborate on choice and take control Agency often has prefential rates with partners
Keep control of 1st party data	All data controlled in-house Requires tech and data-segmentation skills in-house	Data ownership controlled by contracts Agency may operate client owned DMP
Private marketplace deals	Negotiated by in-house team, using advertiser's relationships and scale	Access to agency traders and group scale for better access, pricing, quality
Programmatic products	Advertiser may use programmatic products like Re-targeting or audience networks and work on the transparency with them to understand how they operate and how they benchmark	Advertiser may use programmatic products like Re-targeting or audience networks. agency works on transparency with them to understand how they operate and benchmarks them
Qualified staff	Internal hiring can often be difficult, not every advertiser is attractive for scarce programmatic talent Small teams (below five staff) often not stable Local market (media partner) knowledge needed for each key market to maximise value of Private Marketplace Deals	Larger programmatic teams offer ease and flexibility of access to staff with cross-industry and cross-country knowledge base Often larger talent pool Can provide local teams in each relevant market with deep market knowledge and publisher clout
Internal integration	Best access to internal resources (data) and stakeholders across the business	Need leadership and champions on client side to connect to all resources and stakeholders
Market knowledge	Deep knowledge of the advertiser's business	Broad knowledge of the (programmatic) media industry, cross-industry, cross-market learnings and best practice
Get outside know-how inside	Risk of siloed in-house view, limited exposure and experimentation with new technologies, approaches etc.	Constant learning across clients, industries, technologies and markets can be made available for clients

If the core business process of an advertiser is customer acquisition online and they can afford a programmatic display team of 5+ people in each core market, then it is viable to take it in-house

For most advertisers, the focus should be to invest into strategic know-how to steer the agency at eye-level and drive the data-driven marketing strategy with all internal stakeholders. In addition, in-house analysts add great value by constantly evaluating performance and challenging the agency team.

If the advertiser owns valuable first party data and wants to make sensitive business or customer data available for programmatic campaigns, the investment in an in-house audience DMP and data management or segmentation capabilities may be a very good decision. Those teams then can build targeting segments that are made available as anonymous cookie lists to an outsourced buying team that does not need to have access to the individual user profile data points to buy ads efficiently.

As the market evolves and advertisers have choice of various operational models, the discussion about in-house versus agency turns into a normal service discussion and can lead to the optimum mix of in-house and outsourced resources.

At the same time, programmatic products that combine data, technology and media inventory in unique ways will also stay very relevant for advertisers and will be components of a successful programmatic setup that leverages all opportunities to drive ROI.

4.3 Moving Beyond the Web: Programmatic TV, Outdoor, Radio

As digital media moves into increased programmatic execution, advertisers ask how they can apply the same principles to other media channels.

Digital Outdoor is already starting to be ad-served, so buying it programmatically is the next logical step. However, the one-to-one relationship of consumer profile and ad-space as in Display advertising will not happen: Outdoor ads can be seen by many people at the same time, most of them will not (want to) be recognised, so using user profiles to drive Outdoor buys will not be the core use-case. However other data signals like weather can influence the value of an outdoor ad. Being able to launch campaigns in minutes and change ads in seconds may make programmatic outdoor interesting for both buyers and seller.

The same is true for Radio, but as music listening moves into one-to-one digital channels with web radio and music streaming services, opportunities for programmatic buying are already becoming available.

TV, however, is the largest advertising channel and the power of video to drive both brand values and actions is unparalleled. 'TV usage is still growing in most markets,' but younger audiences move more and more into non-linear video-watching online, so advertisers want to do integrated screen planning, combining traditional TV with online video, to maximise reach and impact.

Buying TV ads programmatically will happen in three ways

4.3.1 Video Neutral Planning: Strategic Planning Across Channels, In-channel Buying

Many agencies now plan video channel neutral across TV and Online, using reach-optimisation tools that calculate estimated reach across screens. Strategic planning tools allocate budgets to the ecosystems of TV, online video and mobile video, but the media execution continues to live in those closed systems such as the TV panel, online cookies, mobile device IDs and closed ecosystems like Facebook and does not allow user-centric optimisation across screens.

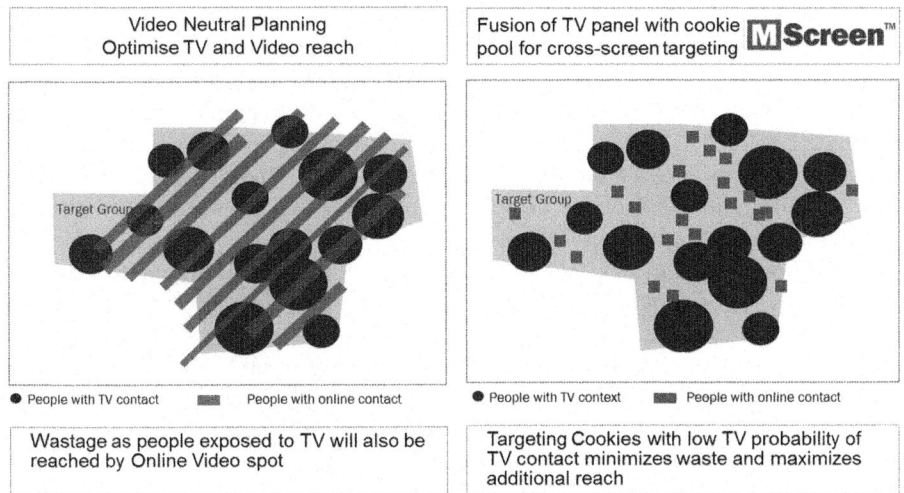

Fig. 3 Cross-screen targeting (Own image)

Online video buying is a challenge as demand for quality video is higher than supply and is fragmented across many publishers – most of which have limited reach. Programmatic buying of video drives significant advantages though cross-publisher frequency capping and applying the same audience targeting data across all vendors.

4.3.2 Combining Online Cookies with TV-Panel for Integrated Optimisation

Video Neutral Planning (Fig. 3, *left*) is only the first step, as we still reach consumers with online video that have been exposed to the TV spot. Integrated tools like MediaCom's MScreen (Fig. 3, *right*) combine the respondent-level data from the TV panel with a large cookie pool such as that in XAXIS. For each cookie, the probability of being exposed to the TV spot is calculated, based on matching of the user-profiles in the TV panel with the user profile of the cookie.

Then only those users not exposed to the TV campaign are exposed to the online video to drive incremental reach without wastage. We have seen significant reach increase at lower cost than through achieving the same reach via TV-only, measured through single-source panels and verified by auditors.

4.3.3 Buying Traditional Linear TV Programmatically

The biggest money pot that ad-tech companies dream of is traditional linear TV. As more and more TV is delivered via Internet Protocol, it can be addressable and traded programmatically. So far the theory.

In reality, programmatic TV buying is still quite far away despite some initial tests.[3] The reason being the complex landscape of content production, distribution rights and advertising sales which is highly profitable, still growing and resisting change. Rights owners sell separate licences for "broadcast" and "video on demand", TV channels package up a linear program and cable or IPTV networks distribute that program, usually not being allowed to alter the package such as sending different TV spots to different households. Therefore the first steps of programmatic selling are very cautious[4]

Where programmatic TV inventory is available, it is often limited in reach and sold with a high premium – making it interesting mainly for high value products sold to small audiences.[5]

In the immediate term, we will see much more innovation happening on the web and mobile rather than in the traditional TV world, which is also limited by regulation. As more video watching is moving into digital non-linear channels, so will advertising budgets, which then will compel the traditional TV industry to adapt to the opportunities of data driven media sales.

5 Conclusion

Programmatic Advertising is not a new media channel, it will be the method to buy media across all ad-served channels. As more and more media moves into digital distribution, Programmatic Advertising will be **the** method to buy media.

However Programmatic Advertising enables more than more efficient media buying: it allows Advertisers to combine data and technology in real-time to move to a Consumer Centric marketing model in which the consumer's needs, her relationship to the brand and her stage in the path to purchase determine which message she will get. Such a Consumer Centric approach will change how advertising is planned and executed, the role of marketing and sales and the role of the supporting agencies fundamentally.

When configuring their own programmatic setup, Advertisers have more choice and control than ever and can combine Programmatic Products with Programmatic Services based on their selection of technology, data, media inventory and specialists to drive maximum Return on Investment. As complexity grows, the role of agencies will grow in importance and they will work in a much more integrated system with the advertiser than ever before.

---

[3] Programmatic TV test ins Australia: https://www.exchangewire.com/blog/2014/11/13/premiering-linear-tv-buying-with-programmatic-tools/.

[4] UK status: https://www.exchangewire.com/blog/2015/04/16/the-dawn-of-programmatic-tv-ad-trading/.

[5] http://adexchanger.com/digital-tv/groupms-bologna-on-the-economics-of-addressable-tv/.

Oliver Gertz is one of the online-media pioneers in Europe. He started his career at Plan.Net media, Germany's first online media specialist that was based on an automated online media buying system with integrated ad server – a very innovative in 1997 and attracting some of the leading online advertisers like Expedia.de, boo.com and letsbuyit.com next to established brand like BMW.

In 2004 he moved as Managing Director to MediaCom, the media agency market leader to grow the digital media service from 14 to 65 staff in the first 3 years and establishing new digital services for major advertisers like P&G, Volkswagen and Deutsche Telekom.

In 2009 he moved into an international role as Managing Director Interaction, EMEA to grow Mediacom's digital offer across 25 markets. Since 2015 he focusses on supporting Mediacom's global clients and defining and rolling out their programmatic strategy.

Deirdre McGlashan Having started during the heady boom days of late 1990s San Francisco, Deirdre McGlashan has been involved in all aspects of digital marketing throughout her career, from strategy to media to technical development and everything in between. As Global Chief Digital Officer for MediaCom, her role is to ensure the full power of technology and interactivity is utilised for our clients, from systems thinking when developing communications solutions, to harnessing and using different types of data, through to delivering specialist excellence in activation, to drive business growth for their brands.

Prior to this role, she was Head of Digital, Global Clients for Aegis Media for 3 years and before that the CEO of wwwins Isobar Greater China (an Aegis Media company), a team of 220 digital specialists across strategy, creative, technology and media across Greater China. In the independent agency space, Deirdre was a Founding Partner at AnalogFolk, a multi-disciplinary marketing communications agency in London that fuses both analogue and digital communications channels.

Deirdre has a vast range of international experience and over the last 10 years has directed on a number of large global accounts including Sainsbury's, Phillips, VW, Procter & Gamble and Sony. She has also delivered award winning work for Adidas, Miller Brewing Company, The Coca-Cola Company and Anheuser-Busch InBev.

Understanding Demand-Side-Platforms

Arno Schäfer and Oliver Weiss

Programmatic Advertising has fundamentally changed the entire spectrum of online advertising. For this form of advertising the technical basics are an essential requirement. The buying decision of every specific ad-impressioxn is made by the algorithm of the Demand-Side-Platforms (DSP) rather than by a person. Technological features are far more important on the demand side than on the supply side.

When it comes to selecting the right partner, it is important to know that there are relevant differences primarily of a technical nature between the various DSP players on the demand side. The US and the European as well as the rest of the online media markets generally have different structures which affect the demand side or the implementation of technology on DSPs.

Very often it is overlooked that also marketers on the publisher side can directly benefit from the technology that was originally developed for the demand side. This means that a marketing company – with or without an Supply-Side-Platform (SSP) – is able to achieve its main objective for its own inventory as an independent purchaser of significantly increasing the eCPM.

The whole industry is rapidly developing and also changing: Using all relevant devices to target advertising consumers or automating the process of technically implementing buying of TV spots will become significantly more important over the next few years and completely change the current way of planning media.

A. Schäfer (✉)
161MEDIA B.V., Erste Brunnenstraße 12, 20459 Hamburg, Germany
e-mail: arno.schaefer@161media.com

O. Weiss
Adform Germany GmbH, Beim Strohhause 31, 20097 Hamburg, Germany
e-mail: oliver.weiss@adform.com

1 Technical Basis

Programmatic Advertising is a very technical issue. The underlying processes and algorithms are extremely complex – also because the technology is constantly being further developed and optimized.

1.1 High Performance Server Platforms

One of the main advantages of using a DSP for Programmatic Advertising is the huge and transparent coverage potential with its connection to all major exchanges and SSPs as well as private marketplaces. Almost all internet users can be reached at the touch of a button – and at a prize determined by the specific and individual requirements of the advertisers.

The demands the system places on the platform are huge, as each inventory request has to be accepted, assessed and processed by the DSP. This happens within fractions of a second. Only those inventory requests promising a high performance are approved for bidding by the opposing party. Each platform in the Programmatic Advertising environment therefore has to be based on extremely complex and lightning-fast hardware and software. It has now become standard procedure to use the latest hardware and highly specialist software. Many providers use cloud solutions that can make global scaling easier and ensure a high degree of consistency.

1.2 Data Processing in Real Time

A core function of Programmatic Advertising is processing data. Irrespective of whether it is 1st or 3rd party data or the simple user history on the platform, all data has one thing in common: It must be available in real-time, which means at the exact moment the bid request on the SSPs is made. The databases that are used for this purpose are highly specialized and are part of the extremely complex hardware and software systems. The amount of data to be processed is currently in the petabyte range but is increasing rapidly. The demand for processing data in real time means that a high data throughput is an essential requirement. Due to the large number of different data sources, the procedure of processing unstructured data places huge demands on the data management systems and especially on the databases. Resulting from the growing demand of big data, modern database models such as NoSQL have been developed. These databases have then been further developed so that new approaches for the underlying operating systems concerning scalability, data, failure safety and redundancy can be added. The demand placed on the databases and the operating system also alters the composition of the hardware. It is now common practice that rotating hard drives are no longer being installed as SSD memories offer quicker throughput rates.

Openness and scalability have never been so indispensable for any of the online marketing technologies like Ad Server or SEM systems, as they are for Programmatic Advertising. A proprietary self-contained solution would not be able to interact with the various specialist providers in the Programmatic Advertising ecosystem. In contrast, due to the high complexity of the Programmatic Advertising market, it is not possible for an individual provider to offer solutions and services for all aspects of Programmatic Advertising. These different systems must be able to cooperate effectively and this only works if they are able to communicate transparently with one another.

The scaling of programmatic systems is generally one of the greatest challenges in the IT sector due to the constantly increasing volume of data used and the simultaneous scalability requirement for the Programmatic Advertising business. It is indispensable for a DSP.

2   Differences Between the DSPs

The most significant differentiator of Programmatic Advertising technologies for advertisers or their agency is the level of service provided by the technical platform. Programmatic Advertising places great and varying demands on the front-end, usability and performance. The media agency can enlist its own employees with the task of operating the platform ("self service") e.g. after initial training and an induction, or as an alternative, they may request that the analysts of the technology provider operate and control the platform as part of the "full service". The third option, known as "half service", describes hybrid forms of complete support service and the licensing of the system. Which form can or should be used often depends on the qualifications held by the employees of the media agency. In the case of major customers, there is an increasing trend towards self-service. This is especially the case in countries that play a pioneering role in Programmatic Advertising such as e.g. the Netherlands in Europe or the USA, as the knowledge of employees there is already at a relatively high level.

DSPs also differ from a technical point of view in terms of how the customer-specific data is integrated, e.g. how existing customer systems are integrated via open interfaces. This is a particularly important criterion for many e-commerce providers. The question closely related to this criterion is to what extent the DSP contains DMP functions, i.e. for processing special, customer-specific data segments ("1st party data"), and allows purchasing and optimization measures to be integrated. This may be an important selection criterion, especially when it comes to creating target group-specific clusters.

In addition to the purely technical features and functions, a distinction can be made between the DSPs primarily by using the following criteria:

- Usable channels: Display, mobile, video, text formats, native advertising, social media
- Usable formats: Only IAB standard formats (IAB 2014) or large formats such as wallpaper, billboards, or customized formats?

Table 1 Differences in the market structure of the online media sector in the USA and Europe (Source: Platform161)

	USA	EU
Inhabitants	310 million (1 country)	500 million (28 countries)
Average test budget	> € 100,000	€ 10,000–50,000
Data protection/Data ownership	Less important	Very important
IP address	Important DSP optimization feature and fixed allocation to the user	Use for optimization purposes usually not allowed and constantly changing
Standard billing types	CPM	CPM/CPC/CPX
Share in available inventory: SSP/Managed inventory	90 %/10 %	30 %/70 %

- Possible purchase methods for the customer: CPM, CpC, CpL, CpA
- Purchasing sources for the customer: SSPs, private exchanges or managed inventory (Picard 2014)

The Programmatic Advertising market is still very much dominated by US providers who have been active in this field for many years and who initially operated with major agency networks from the USA as part of global agreements. However, also European providers have grown massively in the last few years. The extent to which an American or a European provider comes into consideration for a customer must be decided based on the technical features offered, as on closer inspection the two markets in the online media sector are structured very differently (see Table 1).

As an optimization variable, the usage of IP addresses differs particularly between for example Germany and the USA. In Germany it is considered dubious due to the legal uncertainty of using IP adresses and unreliable as the address may also change frequently. An IP address in the USA however is considered to be like a postcode that doesn't change and therefore represents an important optimization variable. The algorithm, for example, is decisively based on the IP address and this criterion has been dropped altogether in Europe. Correspondingly, it would be very unwise to adopt an identical American algorithm in Europe.

It is also important that the media agencies check whether the transparency of the occupied URLs is important to them or not. The customer must be aware of the fact that a lack of transparency of the occupied URLs often also equates to the hidden arbitrage of the provider.

3 Necessary Amendments to a DSP in the Programmatic Advertising Ecosystem

In the Programmatic Advertising ecosystem, the DSP as the main decision-making body forms the technological basis for the purchasing decision and the trade associated therewith. The technology or the algorithm of the DSP makes the decision about demand based on each individual impression available from all available inventories – irrespective of the connected SSP.

And yet, it can be even more precise: In order to realise the full potential, additional components have been developed which some DSP providers are already offering natively and which are described below.

3.1 Data Management Platform

Advertisers and agencies can use the Data Management Platform (DMP) to gain access to advertising-related target groups. The DMP meets multiple requirements such as recording and storing of specific user information, compiling this information into useful and relevant user groups, enabling these groups to be used in targeting and providing the data for the bid calculation.

The qualitative assessment factors of a DMP are real-time capability and simple integration. As significant coverage and quality losses are to be expected from separating the DMP and DSP systems, it is important that the systems of both components are identical. Independent studies show that coverage losses from user profiles of 20–40 % occur if DSP and DMP technology are used from different providers. (Demas 2014)

3.2 Cross-Channel Tracking Including Mobile Channels

A holistic view, i.e. the analysis and enhancement of all marketing activities and all channels is worthwhile for every media agency. This objective can be achieved in the digital world by using a central platform for managing activities. The prerequisite for this is that the system concerned offers technical functions for buying, processing and optimizing different channels, such as display, video, social and mobile. This is executed via special providers and is not necessarily part of a DSP portfolio.

3.3 Conversion Attribution/Customer Journey

Conversion attribution describes the holistic view of user activities in terms of how these activities interact with the customer brand. The basis for conversion attribution is formed by comprehensive tracking of user interactions across different advertising channels and using the customer website. Different parameters are also available for conducting a qualitative assessment of the user interaction such as e.g. advertising size, recency, frequency of contact or the type of interaction (view or click). The user's action on the customer website is always used as the entry point into the user's holistic view and for attribution purposes. Based on this event all interactions of the user with the brand on all available marketing channels are drawn from the history and weighted differently in accordance with the distribution parameters. The objective is to attribute the user's action to the different media channels in real-time and to achieve fair compensation and a correct optimiation at the time of the user interaction.

3.4 Interfaces

Programmatic Advertising often involves customer technologies and constantly used data that must be integrated into the Programmatic Advertising technology. Self-contained Programmatic Advertising systems would make this impossible and would force the customer to always use Programmatic Advertising technology "in a closed space", i.e. separate from its other data and systems. The solution: Application Programming Interfaces (APIs) can be used to link selected parts of external software. APIs are usually used to transfer data and information. The functions of a software solution can be specifically enhanced by adding external technologies by means of open interfaces. Examples of integration and optimization using APIs are first party data, third party data, tag management solutions, ad safety solutions, ad visibility and privacy providers.

3.5 Dynamic Creatives

In order to ensure increased usage of the recorded data and generate an enhanced ROI, it is necessary to optimize media buying and address users with as much accuracy as possible. Each user has different interests and a different purchasing history. Therefore, it is logical to address the user based on his/her profile. Dynamic creatives generated in real-time can be used to display the product statement that is precisely suited to the user; a process that is particularly relevant in re-targeting.

3.6 Tag Management

Practice shows that nowadays each advertising agency employs a number of external technologies on its website. Major e-commerce providers in Germany as well as US, for example, are using partly more than 30 systems for enhancing the way it addresses its users. In order to enhance the speed of the website and to obtain complete control of all the different technology and media partners involved, it is becoming more and more important to use intelligent technical solutions to manage and control these partners.

4 Significance of the SSP Connection

The connection of the demand side with the supply side, i.e. the connection of an SSP with a DSP, can be installed within a few weeks. Most DSP providers are now connected to all relevant SSPs per se, which means that there is no longer any distinguishing feature.

The basis for integrating both sides is the open RTB specification that was developed by Realtime Bidding Project, a suborganisation of the IAB. This describes the basis for automated trading with digital advertising between the market partners. There are differences in the scope of user and inventory information transferred from each SSP to each DSP. The various SSP providers also differ in the additional functions they have to offer. A well-known function that not all SSPs offer is the option of being able to create a private network (Private Marketplace) which provides advertisers or agencies with access to specified inventory under fixed terms and conditions. The basis for setting up this kind of network is the "Deal ID" which is produced within the SSP and is implemented by the business partner into his DSP. The Deal ID therefore acts as a unique identifier and interconnects the systems for the specific business relationship.

In simple terms, the basic function of an SSP is to move the inventory from the warehouse to a shop window. In correspondence, the inventory is much more effectively categorized which results in an increase of the requested inventory volume.

The concrete decision on what is in demand and how much is required must by definition be made on the demand side. In line with the campaign objective or brand protection, the subjective and objective quality of the inventory is hugely important for the DSP. The challenge for the purchasing side is that sometimes different approaches, e.g. floor prizes (minimum bids) or merely a hidden URL transmission (lack of transparency across the actual domain on which the creative banners will be supplied), are being pursued for the individual SSPs. The DSP can support the process of ensuring consistent transparency across the different SSPs by adding additional functions and data.

5 Excursus: DSP Use for Publishers

It is an interesting fact that publishers can also benefit considerably from the DSP technology. By using this technology, the publisher can continue to strengthen his position towards the demand side in the current ecosystem. The question for the publisher therefore is not whether he can use Programmatic Advertising and the potential for standardization or optimization to increase its effective thousand contact prize (eCPM) and utilisation of inventory, but how he can do this. The marketer literally purchases his own inventory while managing the terms and conditions of the bid at the same time. By doing this, he can generate downward pricing pressure on the eCPM of the inventory to be sold and therefore counteract a

Table 2 Benefits of using DSP technology for the media and marketers (Source: 161MEDIA)

Increasing the eCPM	Monetizing the first party data	Enhancing purchasing/revenue sources
Enhancing yield management with efficient "buying" or supplying their own inventory Expanding the product range to include CPC and conversion optimization (CPX) Reducing their own resources (sales, handling) Internal competition for the right impression increases the eCPM (automated vs. own sales)	Segmentation, management and activation of data offer increased control over inventory Adding DSP functions such as re-targeting and segmentation increases the value of the own inventory Generating a higher demand for their own inventory by increasing the product range	Access to other inventory (SSP/Managed inventory) Extending the campaigns and thereby increasing the revenue sources

drop in the eCPM and also specifically enhance the midtail inventory. Table 2 summarizes the benefits for the media and marketers.

As the example below will show, the Netherlands and the marketer Telegraaf in particular are the worldwide leader when it comes to optimizing their own inventory in this sector.

Telegraaf, the largest online publisher in the Netherlands, has been putting all of its inventory on the SSP "Rubicon Project" for many years. The Telegraaf Media Group (TMG) uses a "private DSP bidder" on its own inventory as a purchaser in competition with external purchasers such as the other trading desks, e.g. Xaxis or Cadreon. By using this private bidder developed by the technical provider Platform161, the inventory can be supplied much more efficiently. This means, for example, that competition for inventory and supply efficiency is made: the Telegraaf was able to increase its eCPM by more than 50 %. The Telegraaf also uses DSP technology to offer its advertising customers additional services such as re-targeting, audiencing and CPC bookings without involving external service, which in turn also increases the demand for its own inventory and also internalizes the arbitrage margins of these external providers. (van der Meij 2013)

Many technology providers in the Programmatic Advertising environment either specialize their product range on the demand or supply side. This ensures that the provider focuses on the financial and functional needs of the customer.

In contrast to this there are providers who offer solutions for both sides of the market. Conflicts of interest may result from this configuration. The demand side expects a high ROI on its campaign budgets as a result of using the DSP technology, while the supply side is aiming to generate the maximum eCPM from its inventory. If the technology provider also offers self-managed inventory on its own exchange platform, this may cause an even greater dilemma.

6 Challenges

One of the ways in which Programmatic Advertising is fundamentally different from conventional online media planning is that the marketer no longer remains on the periphery and instead addresses the right user at the right time and on the right device. Under certain criteria, it is irrelevant where the user that delivers the best result for the brand is found, i.e. in which environment or under which URL, as long as the environment is considered "brand safe" by advertising agencies and the campaign objectives are achieved. Efficiency is one of the major benefits of Programmatic Advertising technology.

Performance campaigns were initially the main drivers of Programmatic Advertising as, at first glance, the added value from re-targeting or prospecting in the performance sector in particular is the highest. This means that the CPL or CPO can be significantly and extensively reduced especially for e-commerce providers (MJ 2014). This development has caused many customers to quickly rethink their position (eMarketer 2013) as the added value of Programmatic Advertising and the direct real-time optimization are immediately apparent and the success becomes evident very quickly.

This also applies to Programmatic Advertising branding campaigns. However, it is still not in focus of many marketing departments (von Rauchhaupt 2014). Raising the awareness especially of brand marketing managers for this highly-efficient and profitable online media instrument is without doubt one of the main challenges of Programmatic Advertising.

And despite of all the obvious benefits that Programmatic Advertising has to offer, there are still a few important aspects that need to be identified in order to know which solutions will have to be developed in the future.

6.1 Automated Systems Cannot Completely Replace Human Optimizations

Back when the steam engine was invented, critics infected humanity with the fear that the human work force would no longer be required due to the popularity of this revolutionary instrument and that it would make many workers unemployed. Many critics still believe the same about Programmatic Advertising. However, the steam engine reduced heavy physical work and helped to make tasks intellectually stimulating, rather than causing mass unemployment. Like the steam engine Programmatic Advertising offers impressive benefits for companies. Tasks that are simple, but prone to errors, are solved. Time-consuming, repetitive and complex processes are automated and general operational savings are made. The requirements for working in Programmatic Advertising will of course change: analytical skills and system expertise will become more important, as the constant studying of campaigns and the development of strategies offer the finishing touch towards achieving perfection with these campaigns. Professional flexibility and the desire to develop further are character traits that distinguish a modern

Programmatic Advertising worker. This results in workers becoming more and more professional and leads to companies achieving significant growth. It is certain that the development of Programmatic Advertising will increase rapidly. And of course this increasingly complex technology will have to be managed, monitored and refined by real people in the future.

6.2 Breaking Down Silos of Thought

Programmatic Advertising means networking of channels and integration of systems. This implies that the emerging thought is broken down into silos (this equates to media channels and departments). In the future, there will without a doubt be no more advertising agencies structured into individual departments that compete against each other to acquire new customers. There will also be no system developers who only create solutions for their own departments. Central areas of responsibility, complete transparency and a holistic approach to assessing and responding to advertising will determine marketing activities.

Unfortunately, many market players still do not take full advantage of the technical possibilities available. In Europe, for instance, Germany or Spain – compared to the Netherlands or the UK – is still a little more conservative about this new way of advertising. On the one hand of course only the available inventory coming from a publisher is made available for Programmatic Advertising., On the other hand the inventory is only handed over anonymously or at inflated floor prizes in order to protect the "premium shares" of the marketer. The ever advancing and sophisticated technology however will easily convince more and more marketing decision-makers to enter this marketing field within the next few years. The increasing generational change in decision-making functions and the market development to digital media will support the above.

7 Conclusion and Outlook

The growth rates of Programmatic Advertising have been extremely dynamic in the last years. The next steps will be to focus on the development of the technical connections. Also there will be an increased acceptance of market players on the supply and also on the demand side. Some issues are mentioned here.

An important topic for all participants is the user tracking across all devices. A user for example researches a product during the day and buys it in the evening sitting at home on his sofa probably using a different device. It used to be impossible to track such a device hopping user with cookie-based tracking techniques, the main tracking foundation for years. In the future, Programmatic Advertising campaigns will be managed using different digital channels and it is already starting to be possible to reliably identify users even if they work on different devices. The advertising budget will therefore be managed more purposefully and accurately and this will have a positive impact on the user's experience.

Considerably more often first party data will be used for managing Programmatic Advertising campaigns. Media agencies will become more and more professional using advertising budgets more and more efficiently than they already do today. The big players are already developing their own data strategies. The quantity and also the quality of data that is not collected by the brand owner but is used by him or her will increase.

All digital advertising channels will benefit from Programmatic Advertising. The first successful campaigns for extending television and radio campaigns have already been implemented. Platform161 launched the first European Digital Programmatic Outdoor and Out-of-home campaign in December 2014 (www.platform161.com). Such processes will be standardized in the future and they will be integrated into standard media planning. It will be possible to integrate new marketing channels that have not been automated up to now. This will include Programmatic Advertising on television.

The future of media planning is based on the technical development of instruments for managing and processing available data. This will apply to branding and performance campaigns regardless of the media type: The technology for linking campaigns and also for campaign transparency is in place and will continue to improve. This technology will in the future be used comprehensively across all channels.

Programmatic Advertising is more than ready to take off.

Bibliography

Demas, B. (2014). The great DMP debate, ad exchanger. http://www.adexchanger.com/data-exchanges/the-great-dmp-debate/. Accessed 15 Apr 2014.
eMarketer. (2013). Nearly one-fifth of US display spending will be automated this year. http://www.emarketer.com/Article/Nearly-One-fifth-of-US-Display-Spending-Will-Automated-This-Year/1010156. Accessed 15 Apr 2014.
Iab. (2014). Universal ad package. http://www.iab.net/guidelines/508676/508767/UAP. Accessed 15 Apr 2014.
MJ. (2014). Case study: Wie Unilever durch RTB den ROI um 400 % steigern. (How Unilever increases the ROI by 400 % with RTB.) Spacebidder. http://imb.donau-uni.ac.at/spacebidder/case-study-steigern-sie-wie-unilever-mit-rtb-ihren-roi-um-400/. Accessed 15 Apr 2014.
Peterson, T. (2014). Digital radio gets real-time with zaxis. Adweek. http://www.adweek.com/news/technology/digital-radio-gets-real-time-xaxis-146698. Accessed 15 Apr 2014.
Picard, E. (2014). The difference between programmatic RTB and direct. Ad exchanger. http://www.adexchanger.com/data-driven-thinking/the-difference-between-programmatic-rtb-and-direct/. Accessed 15 Apr 2014.
van der Meij, M. (2013). Real-time with De Telegraaf's van der Meij. Mediapost. http://www.mediapost.com/publications/article/196266/. Accessed 15 Apr 2014.
von Rauchhaupt, J. (2014). Programmatic for brands? adzine. http://www.adzine.de/de/site/artikel/9701/adtrading-rtb/2014/01/programmatic-for-brands. Accessed 15 Apr 2014.

Arno Schäfer has been working in the digital industry for more than 17 years. He started his career as a consultant at Roland Berger & Partner. After co-founding and establishing the mobile marketing agency MindMatics in Munich, he became Managing Director of MediaCom Interaction GmbH in Dusseldorf in 2005. From 2010 to 2011 he was Partner and Managing Director at Performance Media Deutschland GmbH in Hamburg. In 2012, he became Managing Director of Digital Response GmbH, which was sold to ClickDistrict GmbH in 2013 a company dedicated to Programmatic Advertising. Schäfer became CEO of 161MEDIA B.V. in 2013 and CEO of Platform161 B.V. in 2015 with headquarters in Amsterdam and with 80 employees in 12 countries. He lives in Hamburg.

Oliver Weiss has more than 15 years experience in senior management and consulting for digital companies. He used to be Country Manager DACH at Platform161. Before that he was Sales Director of the Cloud Marketing Platform Turn and was responsible for managing the market launch of the company in Germany. His previous career steps include working as the General Manager Europe at Facilitate Digital, as the Head of Atlas Advertiser Suite at Microsoft and as a Consultant at DoubleClick. He also worked for pilot 1/0 and for Pixelpark. At Adform he is now responsible for international product coordination and client partnerships.

Granularity Creates Added Value for Every Objective

Arndt Groth and Viktor Zawadzki

Since becoming mainstream around 2010, at least in the United States (http://www.sfgate.com/technology/businessinsider/article/The-Rise-Of-Real-Time-Bidding-Is-The-Biggest-2463215.php), Programmatic Advertising has grown from an isolated point solution of buying cheap reach of unsold inventory into an integrated concept for online marketing.

The integration of advertising measures along the marketing funnel (http://www.mckinsey.com/insights/marketing_sales/the_consumer_decision_journey) or the hourglass model (http://www.hotelmarketingstrategies.com/marketing-an-hour glass-2353/) requires various strategies and tactics that have to be developed in a coordinated manner, continually optimized and supplied with data. This data is aggregated and analyzed in real time and then fed back into a media buying and campaign optimization system.

Programmatic Advertising campaigns require a precise definition of objectives and planning. The reason for switching from classic media planning to Programmatic Advertising is the far greater efficiency of such marketing measures, as well as internal and external workflows, and scale.

This article provides an overview of the components and a presentation of the critical mechanisms for successful media planning in Programmatic Advertising.

A. Groth (✉)
PubliGroupe Ltd, Avenue Mon-Repos 22, 1002, Lausanne, Switzerland
e-mail: agroth@publigroupe.com

V. Zawadzki
Spree7 GmbH, Dorothenstr. 35, 10117, Berlin, Germany
e-mail: viktor.zawadzki@spree7.com

1 Always Act "Programmatic First" or "Programmatic Media Planning"

The programmatic campaign process takes a somewhat different approach than classical media planning. In recent years, rigid media plans have been increasingly loosened up and redefined. What advertisers have been successfully practicing when it comes to search machine marketing – namely the continuous delivery and ongoing optimization aimed at predefined campaign objectives, most typically return on marketing investment – is relatively slowly, but surely, gaining ground in other online channels. Campaigns are no longer planned in fixed-interval flights, but are run continuously on an ongoing basis.

The fundamental principles of marketing remain valid despite automated workflows and campaign optimization with algorithms and big data.

As the name implies, the major advantage of Programmatic Advertising is the optimization of marketing measures in a systematic and automatized way. Ad space is no longer purchased in packages, per thousand ad impressions (cpm), but on an individual basis. The value of each ad impression is determined and traded in mostly about 50 ms.

As in classic media planning, campaign structure planning in Programmatic Advertising occurs along a sales funnel.[1]

2 The True Campaign Objectives

More and more marketers have discovered their interest in Programmatic Advertising and become much more willing to use, or rather to experiment with, this technology. They ask questions like: How can advertisers make the best use of this technology? How do I plan a successful campaign or, even better, how do I use Programmatic Advertising to boost the success of my online marketing activities? It is highly inadvisable to regard Programmatic Advertising as just another inventory source and make buys as one would with an ad network.

The preliminary work in Programmatic Advertising consists in determining the campaign objectives. Precise definition is the prerequisite for a successful Programmatic Advertising campaign. What may sound relatively simple in fact requires an investment of time and resources to determine distinct key performance indicators to guide campaign optimization efforts.

To this end, mainly the following questions have to be answered:

1. What do I want to achieve with the campaign?
2. How do I measure success?
3. Which measure is most effective in terms of optimization?

---

[1] Refers to: http://www.exchangewire.com/blog/2013/11/29/video-iponwebs-boris-mouzykantskii-on-unpacking-the-black-box-ats-london-2013/.

PROGRAMMATIC-ADVERTISING ALONG THE CUSTOMER CONTACT FUNNEL:

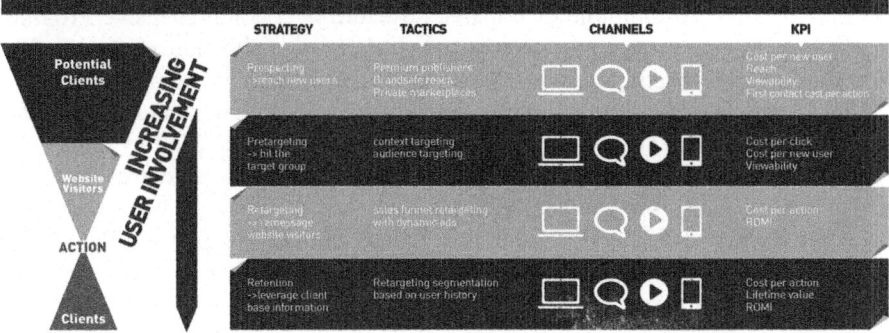

Fig. 1 Full funnel marketing (authors' own compilation)

The work with advertising clients and agencies is characterized by two extremes among clients: performance-driven clients who are purely driven by the cost per order (CPO) targets, linking each individual ad impression to a direct sale on their website, and the classical advertising clients who consider more indirect goals or branding effects like target group contact, reach and click rate of a campaign as factors of success.

The main focus in defining the objectives of a Programmatic Advertising campaign consists of identifying the true goals of an advertising client and hence to eliminate the pseudo- and placeholder goals.

The marketing funnel in Fig. 1 shows the starting point for the development of a comprehensive Programmatic Advertising campaign. The funnel and the user segmentation can be expanded in depth and breadth as desired in order to exploit further areas of potential.

The primary objective of most global marketing measures is to maximize the absolute contribution margin. Simply put, this means increasing the number of sales and revenues and lowering the acquisition or activation costs per customer.

These overarching objectives can be divided into more tactical sub-objectives. Potential sub-objectives might be:

1. Increasing the number of new customers
2. Boosting revenues/customer lifetime revenue per existing customer
3. Lowering acquisition costs per new customer

4. Decreasing activation costs per existing customer
5. Increasing the number of sales per existing customer
6. Increase reach in the target group

3 Full Funnel Marketing

Figure 1 shows the simplified campaign structure of a comprehensive Programmatic Advertising campaign encompassing all parts of the sales funnel – from branding and prospecting to retargeting and marketing to existing customers.

The planning and execution of such comprehensive campaigns in Programmatic Advertising requires an integral analysis of the marketing and product setup.

The campaign structure should be broken down into modules corresponding to the positions along the marketing funnel and target group. In each module, a strategic approach is defined, then the online marketing channel, the tactical positioning, the optimization objective or KPI and corresponding user segment management are chosen accordingly.

3.1 Prospecting or Branding

In the upper part of the marketing funnel, the users we want to address are far from the point of making a purchase. KPIs for campaigns with the goal of filling the wide upper funnel focus on such goals as increase of new website visitors, or GRP (gross rating points), the standard measure familiar from TV marketing. As a number of providers have already made it possible to measure delivery by visible ad impressions, optimization on viewed ad impressions has become much easier to achieve. The industry standard of a viewed impression is fulfilled as soon as more than 50 % of a banner is viewed for at least 1 s.

As in classical media, along with reach and visibility, the advertising environment or the "placement" is a decisive factor when it comes to high advertising impact. The criteria for the selection of high-quality environments are relatively imprecisely defined in the industry. The qualitative criteria of premium environments, such as the proportion of attractive target groups within the overall audience and a high visibility of the advertising media, can be partly or even fully overridden by factors like viewability, quality management tools and external audience data. Nevertheless, the technical tools only maximize the impact of a campaign. Environments that users perceive as high quality due to the content and credibility of the medium mostly achieve much more favourable results.

In order to achieve the marketing objective of increasing the number of users, which visit the website for the first time, recommended tactics include advertising measures on a selection of premium sites purchased programmatically, which are especially available via private marketplace deals, the increase of reach with a set minimum level of quality, using quality and brand safety tools and the standard negative retargeting approach, the exclusion of users who already had contact with the website.

Due to their high degree of viewability and effectiveness, video advertising media, rich media display and the Facebook Newsfeed are especially conducive to achieving high-level advertising impact, while standard display advertising media offer a cost-effective increase of reach.

3.2 Predictive Targeting

The goal of predictive targeting consists in the acquisition of new users, through employment of targeting tools, which predict a high probability of users' interest in the advertised product in order to minimize divergence loss. By exclusively restricting the campaign to users who fit the target group – whether this happens based on a consideration of already existing customers, on market research, or alternative methods – the efficiency of the budget can be maximized. The probability of a user's behaviour, such as the probability of him purchasing products on websites of the advertiser, which this user has not yet visited, can be determined from information about the user's technical characteristics, the context of visited sites and his anonymous behavioural data derived from external sources. All online campaigns are modulated according to such calculations.

3.3 Retargeting

Which data is actually the most effective in Programmatic Advertising? It is the advertiser's own – so-called 1st-party data. The structure of a classic retargeting campaign follows the lines of the sales funnel. Thus, retargeting – the addressing of previous visitors outside the actual website– became established some years ago as a fixed component of the online marketing mix.

3.4 Marketing to Existing Customers

As shown in Fig. 1, marketing does not end with the successful conversion of a customer. A successful customer acquisition marks the beginning of further cross- and up-selling measures using classical CRM measures such as e-mail marketing, as well as other online marketing channels through which existing customers can be retargeted. In turn, this further segmentation of users along the funnel boosts marketing efficiency.

The general conversion of this part of the funnel into specific marketing measures may seem simple in principle, but can become highly complex in terms of implementation, depending on how these measures are structured. A simple way to begin addressing existing customers is the definition of a "buyer" segment for retargeting and individual addressing via advertising banners or videos based on their buying history. Further segmentation of users can be achieved by splitting up groups according to their behaviour and their reaction to other CRM measures, such as newsletter opening, shopping cart value or regularity of website visits.

The setup possibilities for the utilization of customer data, so-called 1st-party data, ranges from a simple pixel based synchronisation of CRM newsletters and segmentation of users according to their behaviour, up to the complete synchronization of the customer data warehouse based on online and offline customer data with all online marketing channels.

4 Campaign Conception

In order to continue with a classical setup, tactical objectives have to be translated into operational measures. These measures can be defined in terms of several dimensions:

- Campaign KPIs/goals
- Channels and formats (display, mobile, video, social, content, print)
- Devices (desktop, mobile, tablet, TV, (wearables))
- Inventory concept (short-tail and long-tail inventory, brand safety)
- Targeting concept (data, context, technology)
- Creative concept (classical and dynamic advertising media, display, video, social, mobile)
- On site conversion funnel
- Attribution/tracking concept (pixel map, attribution model)
- Optimization planning
- Project plan/stakeholder definition
- Implementation roadmap

The planning and conception phase for Programmatic Advertising are at least as important as the implementation and management of a campaign.

The above-cited operational measures can be structured according to the solution map (see Fig. 2).

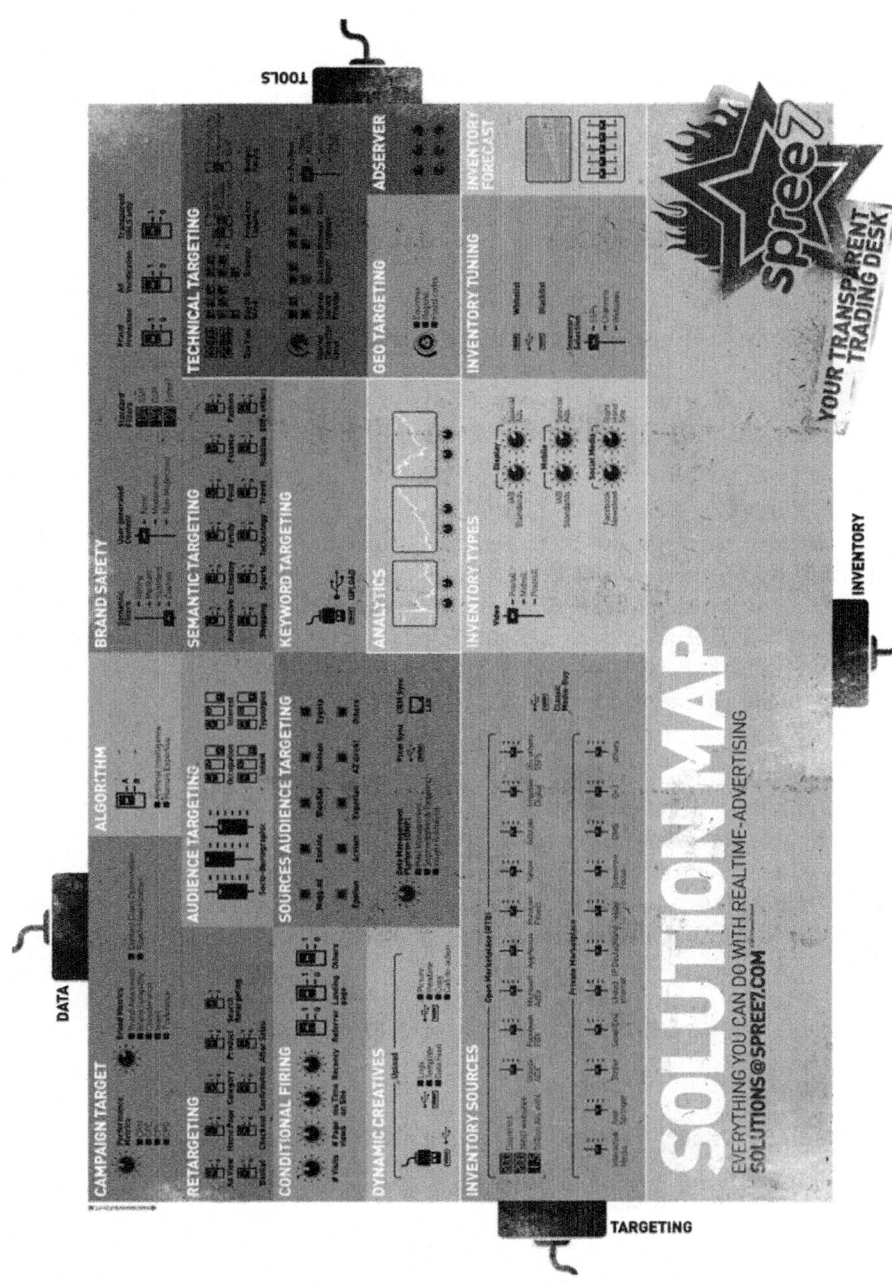

Fig. 2 Solution map – everything you can do with programmatic advertising (Source: http://www.spree7.com/solutionmap)

4.1 Campaign KPIs

Direct campaign objectives consist of cost-based direct goals such as cost per click, cost per lead or sale, as well as CPMs in the target group. Indirect campaign objectives involve brand metrics, which are measured using tools like surveys or market studies. Surveys are also used simultaneous to an ongoing campaign to measure how the perception of a brand changes over the course of the campaign.

The described key benefits can only be attained if Programmatic Advertising is regarded as a framework for complete online marketing and all activities are synchronized accordingly.

4.2 Channels and Advertising Formats

Which online channels should be used for a brand or product? In Programmatic Advertising, the following channels are presently available.

4.2.1 Display
This category involves advertising banners shown on websites. In somewhat simplified terms, these can be divided into standard formats and special formats. A detailed overview of all standardized banner formats can be found on the Interactive Advertising Bureau website.[2] The display channel comes into play for all measures along the marketing funnel. More and more rich media formats become traded programmatically.

4.2.2 Mobile
The mobile channel encompasses all mobile devices, although the distinction between tablets and mobile phones is becoming increasingly blurred. Like for the display channel, advertising banner formats for mobile devices have been defined by the Interactive Advertising Bureau and the Mobile Marketing Association.[3] Accordingly, there is a distinction between standard and special formats.[4] The existence of a mobile-optimized website or a mobile app, however is prerequisite for the effective use of mobile marketing. Mobile campaigns require additional cookieless tracking methods in order to track actions in applications.

4.2.3 Video
The available formats for video are InStream videos, that is, pre-, mid- and post-rolls and in-banner videos, which can be integrated into nearly every standard and special advertising format.

---

[2] http://www.iab.net/guidelines/508676/508767/displayguidelines.

[3] http://www.iab.net/guidelines/508676/508767/mobileguidelines.

[4] http://www.iab.net/risingstarsmobile#1.

4.2.4 Social

Social media also allows for advertising in a programmatic context, most notably via Facebook Exchange in the last 24 months.[5] Facebook Exchange (FBX) loses relevance as Facebook develops its own capabilities further. Cross-device tracking, video ads, dynamic retargeting using 1st-party data, more and more 3rd party data are only some features to mention, which make facebook a very attractive advertising channel.

DSPs, like MediaMath and its acquisition Upcast Social, already provide access to the advertising formats of other social networks. For instance, the demand-side platform MediaMath gives users access to Twitter-tailored audiences.[6] Thanks to this connection, retargeting lists from the MediaMath platform are made available for targeting via Twitter ads.

More recent and fast growing social networks like Instagram, Snapchat, Pinterest, Tumblr, WhatsApp or Vine are getting integrated in the standard marketing mix of each and every professional brand.

4.3 Devices

In addition to desktop PCs and laptops, devices on which ad impressions can be purchased, include mobile devices like mobile phones and tablets. More and more ad impressions become available on connected TVs. MediaMath became the first demand side platform, which offered programmatic print advertising. Digital out of home and radio are to mention as devices, which are still a niche.

However, user-based optimization of campaigns and synchronization of advertising measures over devices remains a challenge, at least for the time being.

4.4 Inventory Concept

The central question of most online campaigns remain: Which websites should my ad be shown on?

This does not change in times of Programmatic Advertising. As the user becomes the focal point in Programmatic Advertising, the environment or website on which a user sees an ad, contributes to the marketing effect.

A Programmatic Advertising campaign consists of ads shown on several hundred up to several thousand websites. Programmatic Advertising campaigns genereally can make use of two different purchasing options.

Private marketplaces allow publishers to make certain ad impressions available to a limited number of buyers or exclusively to one buyer for a minimum auction based or a fixed price.

---

[5] https://www.facebook.com/business/products/advanced-ads.

[6] https://open.mediamath.com/apps/twitter-tailored-audience.

Private marketplaces are one possibility, they are convenient for purchases of high-quality Alternatively, ad impressions can be traded programmatically in an open auction, which is probably the closest to a liberal market – auction based and not limited in terms of buyers.

It is recommended to use whitelists and quality targeting when buying in the open marketplace to be able to achieve campaign KPIs. This whitelist then forms the environment for a campaign. When the selection is made automatically by using targeting methods, it should always be reviewed manually to guarantee 100 % certainty of brand-safe sites.

Depending on the product that is being promoted and depending on the campaign goal, it pays off to consult industry- and target group-specific lists of sites.

These whitelists can be expanded or restricted, depending on the performance and taking into account the respective targeting.

4.5 Targeting Concept

Programmatic Advertising offers various targeting options. If a combination of those is used, it should be borne in mind that the use of more than one type of targeting has a very restrictive effect on the delivery of a campaign.

The following list contains a selection of targeting options, which is by no means exhaustive. It is important to remember that for a certain targeting option, such as contextual targeting, there are multiple technical providers operating in the market, each employing different technologies.

1. Technical targeting tends to encompass more classical factors for campaign targeting. These include frequency capping or contact frequency, time targeting, geo-targeting (postal code, state, city, region), criteria like internet service provider, browser, connection speed, operating system, device, ad position (above the fold, below the fold or non-categorized), budget pacing (evenly distributed or as quickly as possible).
2. Generally speaking, semantic or contextual targeting functions on the basis of an analysis of website content. Websites are pooled in individual channels according to their content and those can be used for targeting. A number of targeting providers allow keyword targeting, similar to the Google's Display network. Contextual targeting allows for the automated delivery of advertising media on thematically fitting websites without having to manually target each site individually in advance. Contextual targeting coincides with pre-targeting in the marketing funnel.
3. The use of anonymized user data is a further targeting option in Programmatic Advertising campaigns. A distinction is made between data that can be collected by the advertiser and used for retargeting and analysis purposes, so-called 1st-party data, and 3rd-party data. The latter involving user data traded via data management platforms, among other sources. A third type of data, so-called 2nd-party data, is becoming increasingly important. This type consists

of user data which, in contrast to aggregated 3rd-party data, typically lacks an indication of origin, without a direct connection between the data owner, such as a website, and the data buyer, who can be an advertiser or an inventory marketer. The advantage of 2nd-party data lies in the scalability and exclusivity of the data in the market.[7]

In terms of the data quality, several factors need to be kept in mind, including whether the information consists of inferred data or declared data that users have voluntarily provided on a website.[8] While the actuality of data plays a subordinate role when it comes to demographics – gender and age being factors that change relatively infrequently –interests and data tell a different story with regard to intended actions, so-called "intent data". The time it takes a user to decide to buy a product is typically quite limited, varying between several hours for low-priced online orders up to several months for new cars or major trips.

Audience targeting or the buying of 2nd- and 3rd-party data is recommended for campaigns addressing users in the upper part of the sales funnel. Especially in the case of branding and prospecting campaigns, the use of audience targeting helps reduce divergence losses.

4.6 Creative Concept

The advertising media's importance to a campaign is illustrated by a Comscore study from 2010, which found that the quality of an ad is up to four times more important than the media plan.[9] By combining this study with more recent data[10] on the effect dynamic advertising media have on the click rate and the absolute effectiveness of a campaign, it becomes clear how critical a role the advertising media play as a determining factor of a campaign. Generally speaking, high-quality advertising material can compensate for a bad media plan or a flawed campaign concept, whereas not even the best media planning and the most sophisticated campaign set-up will make up for bad or irrelevant advertising material.

The same holds for conversion funnels and non-optimised landing pages, which make it impossible for users, even those who are highly interested in a product or service, to purchase the desired product or service online.

When it comes to online advertising media, it is very important to design them according to the advertising format and campaign goal. The optimization of advertising media to target groups and the coordination of advertising media with

---

[7] http://www.adweek.com/news/advertising-branding/second-party-data-can-help-brands-get-unique-information-scale-156429.

[8] http://lotame.com/1st-2nd-3rd-party-data-what-does-it-all-mean.

[9] http://www.comscore.com/por/Insights/Press_Releases/2010/10/comScore_ARS_Research_Highlights_Importance_of_Advertising_Creative_in_Building_Brand_Sales.

[10] http://www.thinkwithgoogle.com/products/dynamic-remarketing-for-retail.html.

the user history increases campaign efficiency to an exponential degree. The most well known application is dynamic retargeting, where the advertising medium displays the recommended product on the advertiser's website. Dynamic retargeting is primarily used in display and Facebook channels. In the upper part of the marketing funnel, eye-catching large-format advertising media tend to be used in order to create awareness. The challenge for the campaign planner ultimately lies in quantifying the marketing effect of each ad impression and targeting campaigns accordingly. For example, the combination of large advertising placements with a disproportionately high CPM will lead to a lower effect per advertising euro spent than that achieved by the use of standard advertising media. The quantification of this effect, however, can only be determined with the help of an attribution model – more on that later.

4.7 Tracking Concept and Attribution Model

1. The tracking concept and the attribution model are directly linked to each other, but have to be just as clearly distinguished from one another.

   The prerequisite for planning the tracking concept is the check to determine which kind of tracking pixels can be used and which user information is available that can be utilized for campaign optimization purposes.

   The most advisable approach is the incorporation of a container pixel solution, which does not affect the site performance and which keeps the online marketing division flexible and independent from the website's release cycles.

   The coordination of tracking of Programmatic Advertising campaigns across channels requires the integration of every marketing channel into a global tracking solution.

   Following the rigorous determination of the KPIs during the campaign planning stage, it is equally important to define a tracking concept that measures the campaign's progress as granularly as possible. In a first step, a pixel map should be defined identifying all subpages on which data or event pixels have to be incorporated and all the parameters including information that is to be transmitted via the pixel.

   The individual data pixels are used for the subsequent definition of user segments while event pixels function as conversion or demarcation pixels for the campaign. A simple start setup is the pixel implementation on a site according to the sales funnel from the homepage or landing page visit to the shopping cart and successful purchase. This pixel structure makes it possible to later segment website visitors in accordance to their actions and resulting position in the funnel, thus generating data for the execution of later Programmatic Advertising campaigns.

   The closer the user gets to a successfully completed sale, the higher his value for the advertiser and the higher the bid for an ad impression. This basic setup can be expanded as desired, but this needs to be done systematically. Each segment follows the goal to utilize the information used to better appraise the value of ad impression of a user. A variety of data sources are available, but the most obvious are 1st-party data from search engine marketing. It is advisable to differentiate users according to the keywords by means of which they arrived on the website. For instance, a user who has reached the advertiser's website by entering the brand name as a keyword will have a higher willingness to pay than a user who has entered a very general search term. Other sources are one's own user database, segmentation of users according to referrers, behaviour on the site (including number, frequency or duration of a visit) or 3rd- or 2nd-party data, which is combined with 1st-party data to more precisely determine the value of a user.

2. In recent years, quite a lot has been published about attribution models. Here, we only mention the fundamental characteristics that an attribution model needs to have: all marketing measures across all channels need to be either directly or indirectly recorded and the resulting data should be used in real time, or as close to real time as possible, for campaign optimization purposes. It is imperative that the model measures and attributes all touch points of a user up until the completed sale. Provided the attribution data is not automatically fed back into the DSP, the recommendation is to fit data and event pixels with a conversion event for each engagement of a Programmatic Advertising campaign. As data is the fuel for Programmatic Advertising campaigns, the highest possible number of data points should be utilized for the optimization of a Programmatic Advertising campaign.

The question of defining the attribution window for an event cannot be answered in general terms. It depends on the percentage of users that have converted within a specific period of time. Similarly, the question of post-click and post-view attribution and attribution period should only be considered after a precise data analysis, taking into account all marketing channels that facilitate a meaningful definition and be adjusted regularly.

4.8 Optimization Planning

The KPIs of a Programmatic Advertising campaign can be enhanced by applying optimization methods. On one hand, the potential of these measures depends on the flexibility of the campaign structure, the campaign duration, the campaign budget and the amount of data that is being collected in the course of a campaign as a basis for optimization. On the other hand, a distinction should be made between algorithmically, or automatically, optimized campaigns and manually optimized campaigns. In the case of algorithmically optimized campaigns, manual interventions interfere with the algorithm. It is generally recommended to algorithmically optimize campaigns, provided quantity of data delivers enough significant learnings. The required data amount varies from demand-side platform to demand-side platform.

Manual optimization is recommended for campaigns with a short timeframe, limited budget and a limited amount of data that can be used for optimization purposes.

The two axes along which optimization takes place, are the effectiveness of maximization of ad impressions using advertising media with dynamic content on the one hand and, on the other hand, the efficiency of purchased ad impressions. In concrete terms, this means showing a user with a certain behavioural history an advertising medium with a certain type of content, such as a product, price and benefit, which maximizes his response probability and, at the same time, optimally sets the price for the ad impression of this user for a certain ad space/targeting combination – in other words, maximizing the contribution margin. By combining effectiveness maximization in the advertising medium and efficiency maximization in the buying of ad impressions, the absolute contribution margin is maximized. The underlying consideration is to boost volume, or rather to increase the number of actions to the point that the marginal contribution margin – the contribution margin for an additional action – is zero.[11]

5 Conclusion

With the increasing technical possibilities, the marketing world is not getting any simpler. Marketing experts have the responsibility to ensure that this new world does not become unduly oversimplified. As new technologies allow for a more efficient use of marketing budgets, the aim has to be exploiting those possibilities to the best possible advantage of the company. The campaign structure discussed here can be seen as a basic framework in the development of a successful Programmatic Advertising campaign. The goal of Programmatic Advertising is not to find the perfect campaign setup, but to manage a campaign structure that allows an ongoing evolution and development of marketing measures. Marketing tools are evolving at such a rapid pace that these recommendations should be merely understood as a solid basis for getting started in Programmatic Advertising. Coming full circle, the conclusion can be made that granularity in marketing creates added value for every marketing goal.

Bibliography

http://www.sfgate.com/technology/businessinsider/article/The-Rise-Of-Real-Time-Bidding-Is-The-Biggest-2463215.php

---

[11] http://books.google.de/books?id=xA6W6qZmS78C&dq=umsatzmaximierung&hl=de&source=gbs_navlinks_s (**s.189**). http://books.google.de/books?id=xA6W6qZmS78C&pg=PA185&lpg=PA185&dq=umsatzmaximierung&source=bl&ots=G-XM-wUr8e&sig=p219LQeqho7yJNnqEO-khCPeIDc&hl=de&sa=X&ei=MNFcU_GCAYKitAbVooCgBA&ved=0CFkQ6AEwBA#v=onepage&q=umsatzmaximierung&f=false.

Granularity Creates Added Value for Every Objective

http://www.mckinsey.com/insights/marketing_sales/the_consumer_decision_journey#
http://www.hotelmarketingstrategies.com/marketing-an-hourglass-2353/
http://www.exchangewire.com/blog/2013/11/29/video-iponwebs-boris-mouzykantskii-on-unpacking-the-black-box-ats-london-2013/
http://www.iab.net/guidelines/508676/508767/displayguidelines
http://www.iab.net/guidelines/508676/508767/mobileguidelines
http://www.iab.net/risingstarsmobile#1
https://www.facebook.com/business/products/advanced-ads
https://open.mediamath.com/apps/twitter-tailored-audience
http://www.agof.de/leistungsspektrum-top/
http://www.agof.de/
http://www.adweek.com/news/advertising-branding/second-party-data-can-help-brands-get-unique-information-scale-156429
http://lotame.com/1st-2nd-3rd-party-data-what-does-it-all-mean
http://www.comscore.com/por/Insights/Press_Releases/2010/10/comScore_ARS_Research_Highlights_Importance_of_Advertising_Creative_in_Building_Brand_Sales
http://www.thinkwithgoogle.com/products/dynamic-remarketing-for-retail.html
http://books.google.de/books?id=xA6W6qZmS78C&dq=umsatzmaximierung&hl=de&source=gbs_navlinks_s (s.189)
http://books.google.de/books?id=xA6W6qZmS78C&pg=PA185&lpg=PA185&dq=umsatzmaximierung&source=bl&ots=G-XM-wUr8e&sig=p219LQeqho7yJNnqEO-khCPeIDc&hl=de&sa=X&ei=MNFcU_GCAYKitAbVooCgBA&ved=0CFkQ6AEwBA#v=onepage&q=umsatzmaximierung&f=false

Arndt Groth has held various management-level positions over the past 17 years, at first primarily on the publisher side at the publishing groups Georg von Holtzbrink, Handelsblatt and at InteractiveMedia, then a subsidiary of Axel Springer. Since 1998 he has worked for leading Internet companies and marketing organizations with a digital orientation. The business school graduate was CEO of DoubleClick in Germany, one of the largest providers of online marketing solutions in the area of Internet advertising, which was bought by Google in 2007, and InteractiveMedia, a marketer of digital media. At ePages, the leading European provider of e-commerce shops, Arndt Groth was managing partner and CEO before joining the global marketing network Adconion in 2008. As President for Europe at Adconion, he made significant contributions to the development of one of the world's largest platforms for the distribution and monetization of digital content. Following the successful acquisition and integration of smartclip in 2011, Groth moved in February 2012 to the advisory board of the Adconion Media Group. Since 1 September 2012, the Hamburg native has served as CEO of PubliGroupe, a leading provider of marketing and media sales services headquartered in Switzerland, whose holdings include a key stake in the performance advertising network Zanox.

Viktor Zawadzki (32) CEO at Spree7 GmbH, the trading desk service for the efficient management of digital media in real time. Spree7 is a partnership between PubliGroupe AG (Swisscom AG) and the globally leading programmatic advertising platform provider MediaMath. The economics graduate previously worked for a company associated with Rocket Internet. As head of display marketing and cooperations, he was responsible for display, programmatic advertising, Facebook, mobile, remarketing and online cooperations for the portals eDarling and SHOPAMAN, preceded by numerous years in a similar position for Jamba! GmbH. The online marketing expert lives and works in his adopted hometown of Berlin.

Enhanced Success with Programmatic Social Advertising

Patrick Dawson and Michael Lamb

Efficient markets operate on the principle of investors having access to the best information when making a purchase decision. Programmatic advertising embraces this principle in its allocation of inventory to advertisers. It provides a single view of the consumer and single point of media planning and buying across all digital media channels, and allows marketers to leverage their own data, media, and technology. The social media ecosystem offers enormous value to advertisers as a piece of Programmatic Advertising; it provides an even more contextually targeted advertising solution where the influence of a friend or other trusted source, guides the user along a path to consuming new forms of content.

With the introduction of more targeted social ad formats (e.g. custom audiences, multi-product ads, dynamic product ads, etc.), social media platforms are embracing the approach to real-time bidding. Traditionally, real-time bidding exists as a facet of programmatic advertising allowing buyers to access real-time inventory at scale. With the growing need to deliver more content as part of a holistic user experience, we've begun to see a shift in using Programmatic Advertising across all types of media.

1 Social Media Today

Social network advertising, in comparison to the ecosystem of online display advertising, exhibits considerable differences. Unlike the traditional web, social networks are more like a walled garden – users must sign on using their account

P. Dawson (✉)
MediaMath, Remo House, 310-312 Regent Street, London W1B 3AU, UK
e-mail: patrick@upcastsocial.com

M. Lamb
MediaMath, 1440 Broadway, 21st floor, 10018 New York, NY, USA
e-mail: mlamb@mediamath.com

© Springer International Publishing Switzerland 2016
O. Busch (ed.), *Programmatic Advertising*, Management for Professionals,
DOI 10.1007/978-3-319-25023-6_8

credentials, and their experience is more or less the same for every user. In this virtual construct, people create, share, or exchange information and ideas in communities and networks,[1] much as they do in real life. The ability to connect with people and share news, updates, and multimedia user-generated content has propelled these websites to become the most popular sites worldwide. Namely, Facebook has become the second most popular website[2] globally, with over 1.4 billion users and over 1 million advertisers. As the leading global social media platform, Facebook presents advertisers with the best contextually relevant information to make an informed marketing investment in social advertising. Facebook's Marketing Partner (FMP) community has become the most intelligent way for advertisers to optimise, deploy and manage their Facebook media spend. Facebook approved FMP's provide a stable and robust solution for advertisers – they innovate in lockstep with Facebook to provide advertising efficiency at scale. There are currently nine specialities within the FMP program that are designed to align with the goals of marketers on the platform (i.e. Ad Technology, Media Buying, FBX, Community Management, Content Marketing, Small Business Solutions, Audience Onboarding, Audience Data Providers and Measurement). In Q4 of 2014 Facebook generated 69 % of their ad revenue from mobile and 100 % of total revenues were programmatic.

The future, which has already started to take shape, will see social CRM generating deeper integrations with existing programmatic advertising systems to bring all advertising and media types under one roof. Twitter has proven to be particularly adept at encouraging advertisers to participate in live events – thus further engaging and activating their user base.

Mobile will continue to act as a catalyst for this intelligent, data driven and high performance programmatic model. TV content is providing a great programmatic opportunity for advertisers, allowing for further development within a greater social footprint. Nielsen's Online Campaign Ratings (OCR) tool is a great example of how using Facebook data to quantify the effectiveness of digital display and video ad campaigns can deliver a greater programmatic approach to Realtime Advertising. Facebook paired with OCR data is paving the way for a more intelligent understanding of audience demographics that view online display and video ads. The recent announcement by Facebook that they will open their Messenger chat application (ca. 600 million users) to software developers so that they can add new features and functionalities, will also be expanded to include businesses seeking to communicate with customers that will be able to track packages and make reservations.

Facebook has also added a payment feature to their Messenger app, to allow transactions within the Facebook environment and, as a result, opens up individual purchasing data directly to Facebook. Social today is connected to a large

---

[1] Social Media Roadmaps – Exploring the futures triggered by social media. VTT Technical Research Centre of Finland http://www.vtt.fi/inf/pdf/tiedotteet/2008/T2454.pdf.

[2] http://www.alexa.com/siteinfo/facebook.com.

population's daily interactions across digital and is becoming an increasingly vital source of success for businesses.

1.1 Behavioural Focus

A key characteristic of social networks is the amount of information that the users are willing to share. With traditional websites, analytics can recognise visits and visitors, and approximate a persona based on the actions visitors take on a website. However, the unique makeup of a visitor is ultimately unknown. It is difficult to know, without a doubt, who they are as consumers. This poses challenges for marketers to develop and deliver their value proposition to the right consumer. Many advances in ad tech have been focused around using data from a variety of sources and delivering personalised experiences across the web. However, on social networks, a wealth of information about its users is available, often self-reported by the users themselves. This can be a great opportunity for marketers to take advantage of to target a particular demographic segment or an audience based on self-reported user data.

Additionally, the value of an ad exposure can be amplified by the social context on social networks. It is in our human nature to pay closer attention to the actions taken by those close to us and in our social circles. Generating a buzz and creating a viral marketing campaign are precise examples of how marketers have taken advantage of the nature of social media, and how social context can be well suited for social networks.

1.2 Facebook Advertising Ecosystem

Since its launch, advertising on Facebook has taken many revolutionary and evolutionary forms. Facebook continues to develop new ways to reach people and drive business KPI's that marketers seek, while balancing the interests of users who do not want to be overloaded with advertising. Social networks can offer relevant and targeted advertising that, coupled with the targeting options available with online display advertising, can really create an omni-channel experience for the end user. Over time, this has become more sophisticated with advanced tools and technology, and an ecosystem has been created to accelerate the development of social network advertising.

The major entities within this ecosystem are the social networks (e.g. Facebook, Twitter), who provide the platform, advertising media agencies who plan and execute campaigns on behalf of advertisers, and ad tech businesses and developers who serve the role of an intermediary between the advertisers and Facebook. When Facebook opened doors to its Ads API (Application Programming Interface) in 2011, a slew of tools were developed to help advertisers manage social ad campaigns. The creation of the Facebook API has enabled development of advanced functionalities and empowered advertisers to work more efficiently.

The growth of this ecosystem has allowed for a continuous feedback loop, contributing to the success of social network advertisers. With sophisticated tools allowing for advanced strategies and granular testing, the Facebook API now provides a more intelligent approach to advertising on social networks. This has brought tangible business results to marketers – P&G for example advertiser saw an 86 % decrease in cost per engagement after the adoption of a social ads management tool.[3] In fact efficiency gains are the lifeblood of Facebook Marketing Partners – innovation, agility and skilled servicing continue to drive impressive results for advertisers via Marketing Partners.

In addition to Facebook, other major social networks have followed a similar approach to advertising, yet none have been adopted at the scale of Facebook's FMP program. Of note is Twitter, who provides a similar ecosystem, offering API access to platform partners that provide the tools to manage advertising on Twitter. While advertising on Instagram and Pinterest is still in its infancy, they are also on the road to providing advanced programmatic advertising opportunities. For example, Pinterest's promoted Pins naturally provide e-commerce potential based on their Pin preferences (i.e. Pinboards) and the types of items users have bookmarked and searched for on the site. In fact, Pinterest has been increasingly positioning their platform as a primary indicator of users intent and aspiration, which is undoubtedly very attractive to advertisers. As Pinterest builds their solution, their robust analytics and deep roots as a discovery engine will make them another important player in the social advertising ecosystem.

1.3   Programmatic Social in Action

Reflecting the development of advanced display ad tech and programmatic advertising, Facebook has continued to offer new solutions for advertisers. Many of its innovations revolve around hyper-targeting and the development of custom audiences. These features allow marketers to granularly target (and re-target) users based on their behaviour.

Marketers are now able to sharpen their focus on a target audience and tailor their messages appropriately. This, in addition to the benefits of social context (e.g. the ability to target friends of users who've responded positively to an ad), makes it an ideal tactic for marketers. These features can now be used and optimised to achieve specific objectives, such as generating brand awareness or driving conversions for different types of products and services.

One particular innovation that has come from the Facebook FMP ecosystem is the ability for paid advertising to leverage viral content. Facebook has carefully managed the balance between organic content and paid reach. Enhancing the reach of content that resonates well with the target audience is an efficient way of

---

[3] T1 Social – "**P&G Hair Care brand Aussie achieves 86 % reduction in Cost Per Engagement on Facebook**".

generating exposure. This feature automatically promotes content based on a predefined virality metric has proven to amplify engagement, as seen in a case study example of the ecommerce retailer, Not on the High Street (NOTHS). They saw a 20.5x fold increase in paid reach versus organic reach.[4] Delivering advertising messages in this programmatic manner – at the right time, to the right audience, with the relevant message, has resulted in high response rates previously unseen in online advertising.

According to Tim Ringel, CEO of Paris based agency holding Netbooster, "Facebook has risen to be a real Google competitor when it comes to ROI driven digital campaigns. From an advertiser perspective, Facebook is more then an alternative or merely an addition to Google – it is a new way to interact with audiences in a hyper-targeted way. Personally, my hopes are that our business vision for a channel and ad format consistent to one to one direct consumer communication will be possible on the back of Facebook's data and the technology available in the industry today".

Bridging the gap between social networks and the web has been another trend across social media. Facebook currently offers several solutions: Website Custom Audiences (WCA), Facebook Exchange (FBX), and Dynamic Product Ads (DPA). WCA is a retargeting vehicle, available by placing a cookie on users visiting a particular website. They can then be recognised on Facebook and marketers can target them as a particular audience. Similarly, FBX is a real-time ad exchange for reaching users that visit a particular website using a cookie, who can then be targeted on Facebook. While Facebook DPA allows marketers to automatically promote relevant products from their entire product catalogue across any device. With the addition of Facebook DPA, Facebook has taken a further step towards fully automated ad creation and retargeting. These Facebook technologies allow for more efficient behavioural and intent targeting and are likely to expand across verticals and geographies in the coming months.

Another example of efficient programmatic advertising on social has been the introduction of Partner Categories on Facebook. Facebook has established partnerships with external audience data providers (e.g. Acxiom, Datalogix and Epsilon). Such integration allows for marketers to target people on Facebook based on their behaviours on other websites and offline experiences, including purchases. For example, a CPG brand could target people on Facebook who purchase a specific soft drink brand offline, using data gathered from a grocery retailer customer loyalty card program. The connection between social network advertising with the online display ad ecosystem and offline consumer behaviour is taking place, transforming the industry and how marketers can reach consumers.

Other social networks also provide marketers with advanced ad tech. Twitter has been a prime example of how marketers can take advantage of programmatic advertising, with brands successfully starting conversations around specific real-time events and opportunities. We've seen this with Oreo's successful "You can still dunk in the dark" Tweet during the blackout at Super Bowl 2013. Twitter also

---

[4] T1 Social – "**How can ecommerce retailers achieve winning ROI on Facebook?**".

offers advanced audience targeting through its Twitter Tailored Audiences (TTA), a retargeting technology that leverages 1st and 3rd party data (similar to Facebook's Custom Audiences, and TV conversation targeting which allows brands to reach people who engage with specific TV shows).

The emergence of social mobile apps is also on the rise with companies like Snapchat, Line, and Kik, – all of which are aggressively integrating with other partners (eCommerce players for example) to compete more with established social networks like Facebook and Twitter.

1.4 Power to the People

As users begin to interact with multiple devices it is becoming increasingly important for marketers to understand that digital measurement standards that are based around the cookie are missing an increasingly large part of the measurement picture. Atlas data has revealed that cookie and browser based views of the purchase journey are on average leading to a 26 % overstatement of reach, a 41 % understatement of frequency and at least 21 % of conversions are not being captured Last click attribution models do not include the power of mobile advertising in driving interest for products that are bought on desktop devices where transacting is typically easier. Therefore, existing attribution models are largely inaccurate as they give much of the value to desktop clicks, which are actually driven from other (mobile) devices. Even multi-touch models cannot track mobile or Custom Audiences impressions, since view tags do not work in mobile apps, on custom audience targeting, or on many mobile browsers.

In order to generate greater measurement efficiencies Facebook is working with Nielsen to bring user level reach measurement to market via Online Campaign Ratings (OCR). In Europe, Facebook works with research company GFK among others, to link Facebook with cross-device exposure to GFK's Media Efficiency Panel and with agencies around the world to supply them with cross-device reach and frequency data optimized for use in econometric models.

Facebook is innovating to provide new people-based measurement techniques which are helping sophisticated advertisers to optimise their Facebook paid media activity to better understand its true value.

2 Vision of Programmatic Advertising and Social in the Future

As consumers can be reached across multiple touch points like desktops, smartphones, and tablets, marketers no longer address consumers as a monolithic experience. Programmatic advertising is melding into an omni-channel method of buying and selling ads across search, social, video, television, native, in-app advertising, and more. As MediaMath CEO Joe Zawadzki notes in his book *MathMen*, "Programmatic will, in effect, provide a kind of digital nervous system that coordinates how the world responds to us – the people who live within it – by optimising our experiences every time we find ourselves interacting with many varieties of smart technologies.[5]"

Given the continual growth and investment in ad tech, one thing is for certain – programmatic advertising will continue to become more sophisticated, with social data taking a more defining role in this process. Online advertising will become interdependent with social network platforms, providing the social context in all forms of marketing communication between a brand and consumers. As a result, marketing messages will be delivered with more relevancy, efficiency, and context. The way brands manage communications and relationships with consumers will become more social and personalised in a highly targeted 1:1 marketing approach.

As mobile devices become more interlaced in our lives, mobile will show it's worth as a valuable ad platform. This trend will only further grow by the advances in user tracking technology beyond cookies, as it will provide marketers the ability to understand and pinpoint where their potential audiences are. In particular, a new social CRM will see deeper integrations with online advertising systems and feed into social networking sites. Mobile is a catalyst accelerating this integration – as we no longer spend most of our time using a desktop. More social mobile apps are beginning to emerge that will redefine the social advertising space in the next few years. Apps like Snapchat, Kik, and Line let their users share personal experiences like photos, chat, and phone conversations with their friends and family. Marketers looking to leverage that native data will turn to these platforms for a way to convert their own audiences and mobile app installs.

Our relationship with brands will become hyper-mobile and hyper-targeted by the use of data available from social networks. As marketing budget spent on traditional media outpaces consumer time spent on those channels, mobile and digital will command more attention from marketers – to the tune of over $20 billion USD in potential spending in US alone.[6]

Programmatic advertising is sure to expand beyond its original purpose of selling remnant display inventory. MediaMath's Joe Zawadzki sees the future of Programmatic Advertising as "a single dashboard where brands will be able to track all of the ways they interact with those individual consumers they care about, and who care about them. With sophisticated attribution modelling, advertisers will be able to consolidate a full array of customer interactions across a multitude of channels and devices into a holistic picture of consumer engagement."

3 Conclusion

This holds true for the future of social network advertising as well, where high performance, data-driven programmatic advertising coupled with a holistic platform will require advertisers to think intelligently about how to leverage data from

---

[5] *MathMen*- Joe Zawadzki (expected release date 2015).

[6] Internet Trends – D11 Conference (2013), Mary Meeker and Liang Wu, KPCB.

their customer lifecycle in a social centric environment. As a result, socially enhanced programmatic will become more vertically dynamic and personal while integrating more deeply across multiple devices to improve the efficacy of social campaigns.

Patrick Dawson is responsible for the commercialization of MediaMath's T1 Social and also works very closely with Facebook's Marketing Partnership's team in ensuring that MediaMath's Social Media efforts are aligned with Facebook ecosystem opportunities. Patrick's remit is global. He led the overseas expansion of Upcast Social (T1 Social) into Singapore and Berlin prior to the acquisition by MediaMath in Q4 of 2014 and has been responsible for growing the business together with MediaMath across some of their key global markets and accounts. Patrick brings over 12 years of experience in digital marketing and media. He has built businesses in Japan, Korea, Germany and North America. Most recently Patrick was Managing Director at Facebook Ads API partner TBG Digital North America (acquired by Sprinklr) and led the commercialization and internationalization efforts for Facebook Page Management partner 'Context Optional' (acquired by Efficient Frontier).

Patrick holds an M.A. in Business and Economics from the Trinity College, the University of Dublin. He lives in Dublin and London.

Michael Lamb is responsible for corporate strategy and development, the commercialization of the company's offerings, and for the success of its business partnerships. Michael works closely with the sales, product and operations teams to solidify MediaMath's position as a leader in the industry while identifying and executing upon growth opportunities.

Michael brings nearly 15 years of experience in digital marketing and media, most recently as a partner at McKinsey & Company where his clientele included many of the world's most prominent content owners, publishers, distributors, and advertisers. Michael is a frequent author and speaker on analytic marketing and digital business models for media companies. Earlier in his career, Michael was a co-founder of Poindexter Systems (now [X+1]) alongside Zawadzki.

Michael holds an M.A. in Mathematics, with distinction, from the University of Oxford and a B.A. in Applied Mathematics, cum laude, from Harvard University. He lives in Brooklyn with his wife and three kids and is sure to be the worst poker player on MediaMath's Executive Team.

Programmatic Brand Advertising

Stephan Noller and Fabien Magalon

To make Programmatic Advertising platforms useful for branding campaigns, several additional requirements have to be fulfilled. The core requirements are the delivery of advertisements by target group and ad exposure as well as according to specific branding KPIs within a specific inventory selection, taking into account not only the importance of a brand safe context, but also that Premium Brands (with impactful ad sizes) are scoring high on viewability, today's Programmatic Advertising systems only support these requirements to a certain extent and need to be upgraded accordingly. Furthermore, a special data strategy is needed which has to be aligned to the specific goals of branding campaigns. Additional measures must be put in place to ensure that an adequate level of data protection is preserved when using Programmatic Advertising technology.

1 Preliminary Note

Technologies used in Programmatic Advertising and the associated strategies are all derived from the field of performance advertising and remnant ad sales. Furthermore, the main players have their roots in markets where digital investments are primarily directed to performance campaigns and the media environment is relatively weak. On the other hand, the potential benefit of Programmatic Advertising is obvious for customers who are using advertising for brand development or to launch new products (known as "brand advertising"). After all, it is possible to achieve high reach with good exposure rates within defined target groups using

S. Noller (✉)
nugg.ad AG, Rotherstr. 16, 10245 Berlin, Germany
e-mail: stephan.noller@nugg.ad

F. Magalon
LiveRail, 112 avenue de Wagram, 75017 Paris, France
e-mail: fabien.magalon@liverail.com

Programmatic Advertising on guaranteed premium inventory with high impact ad formats delivering high viewability scores – in other words, this is exactly what brand advertisers have been offered in the past within television environments. The following paper aims at identifying the requirements of Programmatic Advertising for brand advertisers, in order to define the ideal strategy of how you can benefit from Programmatic Advertising technologies as a publisher of online environments or as an agency, when it comes to large-scale budget funds for brand advertising.

2   What Is Programmatic Advertising for Premium Advertising/Branding and Are Specific Strategies Required?

Every single advertising effect has to go through the 'needle eye' of human experience and behavior.[1]

2.1   Definition

Within this context, there are three major characteristics. The advertiser is pursuing different targets for brand advertising than he would for advertising to increase sales. If the primary objective is to increase awareness of or affinity with a brand, or to publicise a new product, the ads are only indirectly aimed at purchasing effects. The direct goal is the "mental click", in other words the consumer is meant to remember the brand shown to him and to make positive associations with it. A possible additional effect is to increase the willingness to buy.[2] As these effects involve attitude changes and memory effects, it is not possible to measure them directly by observing the consumers' behavior. Instead indirect measurement methods are to be used, such as those used in market research. Consumers could therefore be asked as part of a test control group whether they (prompted or unprompted) remember a particular brand for instance. It is also possible to carry out a series of specialist psychometric procedures such as measuring eye movement, activation measurements or the like. Admittedly, the attribution of the effects to the criterion being measured is frequently unreliable and the application of these procedures is complicated.[3]

Brand advertising also attempts to achieve the desired effect by controlling ad exposure. This is because numerous studies have for instance shown that certain memory effects occur only after being exposed to an advertising message a certain number of times (Fig. 1). The exposure level that would be effective for a given target frequently falls within a range of a minimum of three and maximum of eight

---

[1] Rosenstiel, Lutz von & Kirsch, Alexander. (1996). Psychologie der Werbung (Psychology of Advertising), p. 15.
[2] Kotler, Philip & Bliemel, Friedhelm (2001). Marketing Management, p. 934ff.
[3] Schweiger, G. & Schrattenecker, G. (2009). Werbung (Advertising), p. 322ff.

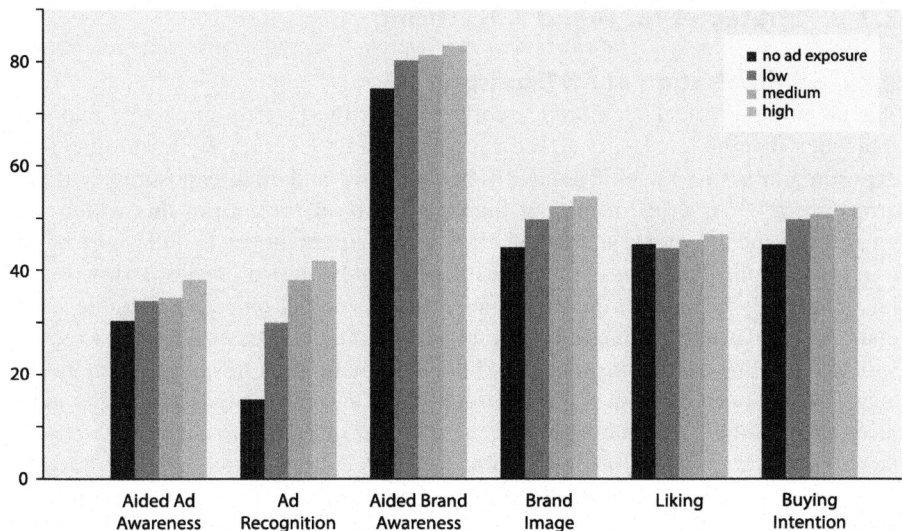

(Source: nugg.ad. Data based on Research & Results 4/2008)

Fig. 1 Influence of exposure on the advertising effect (Source: nugg.ad. Data based on Research & Results 4/2008)

to ten times. However, this should be established in empirical terms depending on the product and the message.

> On the other hand evidence suggests that we could assume threshold values, in other words the fact that a certain level of exposure is necessary in order to achieve any effect at all. On the basis of analysis carried out so far, it is the declining effect curve that has the greatest empirical relevance.[4]

According to this theory, too much exposure may involve the danger of the so-called "boomerang effect", in other words the advertising effect achieved is the opposite to the originally intended.[5]

The third element of brand advertising is the target group. Brand advertising is usually directed at a pre-defined target group, frequently on the basis of a combination of socio-demographic variables. On the whole it is particularly difficult to define the target group profile precisely with new products or brands. For this reason a broader definition is recommended which can be adjusted later based on empirical findings.[6]

---

[4] ibid. p. 302.

[5] Felser, Georg. (2001). Werbe- und Konsumentenpsychologie (Advertising and Consumer Psychology), p. 290.

[6] ibid. p. 379f.

2.2 Strategies for Brand Advertising

2.2.1 Optimisation of Ad Exposure

The advertising effect measured usually follows the pattern across the exposure frequency that the increasing effect starts at a certain level (rarely fewer than three exposures), rises to a peak (frequently between five and eight exposures) and then drops again.[7,8] A series of critical findings can be derived from this which may impact the media strategy: first of all it is important never to fall short of the minimum exposure, because messages that are only shown once or twice usually fall completely flat and are therefore ineffective. Even this objective – simple at first glance – is difficult to achieve in online media strategies because typical exposure spread for online campaigns mostly exhibit a dramatic peak in spread when there is a single exposure. It is not infrequent that 80 % and more of a campaign's target reach is achieved here (in other words the number of people reached).[9] A series of factors specific to the online medium lead to this extreme challenge: the transient nature of cookies and other identifiers is worth a mention at this point, as well as the highly fragmented character of the medium. In practice this is exacerbated further by the universal usage of search engines. The probability that a user will just happen to be directed past a particular page after entering a search term is significantly higher than the chance of accidentally switching on a television or radio channel, or even accidentally opening a printed magazine. Furthermore there are notably more channels/websites dividing the attention of users. This audience fragmentation leads to the issue of insufficient ad exposure of online advertising, especially if it is only delivered through a single website or just a small number of sites. Media planning strategies as well as technology (so-called "frequency boosting") are needed to help tackle the issue.

2.2.2 Measuring the Effect

Indirect approaches are often used to measure the effect. This means attempting to run an advertising campaign using parameters and/or strategies with a proven or assumed connection with the actual target indicator.[10] One reason is that to date it has only been possible to carry out direct measurement of such effects alongside the campaign in very few cases. Indirect measurement can be carried out through post-campaign market research or introducing a media planning strategy aiming to use brand transfer effects from the media brand to the advertised brand. However, all of these indirect methods have two problems in common. Firstly, the link between the

---

[7] Felser, G. (2001). Werbe- und Konsumentenpsychologie (Advertising and Consumer Psychology), p. 387.

[8] Schweiger, G. & Schrattenecker, G. (2009). Werbung (Advertising), p. 302.

[9] A further problem is that parameters operating according to average exposure, such as GRP, often conceal this connection.

[10] Rosenstiel, Lutz von & Kirsch, Alexander (1996). Psychologie der Werbung (Psychology of Advertising), p. 56ff.

indirect criterion and the target criterion is frequently tenuous or generally questionable. Secondly, a strategy like this makes it impossible to optimise the campaign while it is running – a method that is considered to be one huge benefit of the online medium. It is only by using measured parameters (KPIs) alongside the campaign, which can be made available in Programmatic Advertising platforms, that makes live control of branding campaigns possible.

2.2.3 Target Group Reach

When it comes to reaching the intended target group, an approach has become established over the last few years, which was originally developed for planning print media campaigns. This approach attempts to identify the (socio-demographic) characteristics of the users of particular websites or content areas within sites (for instance by conducting surveys). The data is then used to plan online campaigns on these websites. This means for instance that advertising for the target group "families" is delivered across environments known to be used frequently by adults who have children.[11]

This approach is also indirect and it is fairly imprecise when scrutinised more closely, becoming even intensified when the highly fragmented nature of online use is taken into account (which is likely to further amplify as a result of the general phenomenon of "media clutter" across all forms of media[12]). For this reason, so-called targeting processes have been developed, which use a variety of methods to attribute direct (socio-demographic) characteristics to users of online media so that they can be addressed with the right kind of advertising.[13] However, even these systems are not error-free, and as well as data protection-related challenges (see Para. 4), challenges regarding precision of data are frequently encountered. This is intensified by the fact that newer processes are able to identify the properties of online users statistically. This offers huge benefits with regard to the reach of targeting systems (cf. Para. 3.3), but also opens up new questions with regard to the validity of the statistical processes used.

To summarise, it can be said that existing technology in principle can be used to carry out brand advertisement through online media, and to achieve good results, as proven by numerous case studies.[14,15]

---

[11] Schweiger, Günter & Schrattenecker, Gertraud. (2009). Werbung (Advertising), p. 304.

[12] Kardes, Frank R. (2001). Consumer Behavior and Managerial Decision Making, S. 73.

[13] Wikipedia: Targeting (Online-Marketing). URL: http://de.wikipedia.org/wiki/Targeting_(Online-Marketing).

[14] nugg.ad: Case Studies. URL: http://nugg.ad/de/case_studies.html.

[15] Tomorrow Focus Media: Adeffects Digital 2013. URL: http://www.tomorrow-focus-media.de/marktforschung/werbewirkung/info/adeffects-digital-2013-wirkung-von-online-und-mobile-werbeformaten/.

2.3 Challenges in Applying Strategies to Programmatic Advertising

Programmatic advertising means that the process of booking, delivery and fine-tuning online advertising campaigns is automated and mechanised to the greatest possible extent. Many specialists consider that automation and simplification of the process is the decisive factor of Programmatic Advertising: to a greater extent than marketing and bidding functions, which are often never even used. "This highlights a reason as to why we urgently need this development in the market. Media planning has always been a complex business. However, the fragmented use of media renders any efficient media plans, that have been formulated manually, a particularly big challenge that requires considerable time and effort." Kristian Meinken, Managing Director of the media agency pilot Hamburg GmbH & Co. KG.[16]

Implementing the above mentioned strategies and fine-tuning systems in Programmatic Advertising platforms available today is a huge challenge, because the technologies they are based on were generally developed for performance campaigns. This means that the standard does not include frequency exposure control or target group planning – not to mention the option of running and optimising a campaign on the basis of branding parameters/KPIs.

In order to run branding campaigns with up-to-date Programmatic Advertising platforms a whole range of upgrades and enhancements are necessary, and it will not automatically be accomplished by playing some tricks. Programmatic advertising platforms need to extend their data interfaces to include audience provider interfaces, with appropriate planning and control options – ideally including the possibility of creating audience-based forecasts. An exposure control option must be set up, in particular to avoid too little exposure (this is important to note as most technologies have implemented frequency capping which is less of a technical challenge but rather helping to prevent too much exposure).

And finally, the set-up of delivery and optimisation strategies based on branding parameters must be feasible. In other words, targeted delivery based on uplift per view of how aware people are of a certain brand, for instance.

Even if none of the platforms on the market currently provide the full range of features/functionalities required, the goal nevertheless seems achievable. The fact is that the docking points needed for two of the requirements stated already exist within the technology – the only major challenges for providers are likely to be frequency control and forecasting.

---

[16] Meinken, Kristian (in interview with Annette Mattgey): Pilot boss Meinken: "Less exposure, same effect with Programmatic Buying". URL: http://www.lead-digital.de/aktuell/mobile/pilot_chef_meinken_weniger_kontakte_gleiche_wirkung_durch_programmatic_buying.

2.4 Inventory Strategy

The volumes of inventory available on ad exchanges seem to be unlimited: in France, for example, 100 billion ad impressions are for sale programmatically on supply side platforms every month.[17] By gathering all supply sources in one central buying platform, demand side platforms have managed to defragment the supply so efficiently that it seems to be unlimited. However, unlimited supply is a major fallacy as quality supply is limited.

It is essential for a successful branding campaign to ensure running ads exclusively on quality supply. Three key metrics define quality supply in a programmatic environment:

- Brand safety: Ads need to be run within an appropriate context by blacklisting potentially harmful sites.[18]
- Premium context, appropriate to the advertising brand: White listing features need to be defined and set in order to manually validate all inventory sources of the running branding campaign.
- Average viewability: Impression per impression needs to be systematically measured by viewability and the delivery needs to be optimized around this metric.[19]

3 Data Strategy

It is not uncommon that Programmatic Advertising is often directly linked with data-driven online advertising. This is based on the fact that an ad inserted in a Programmatic Advertising system usually has to function independently of the specific content environment in which it is run, without providing additional information. Consequently, an essential feature of conventional online media planning disappears and needs to be substituted by using appropriate data.

3.1 Data Types Along the Funnel

Depending on the goal of an advertising measure as well as the customer's situation, an alternative data strategy needs to be adopted. If the customer is about to buy a product, sales-promoting advertising should be delivered with the aim of sealing the deal. However, if the customer's attention is to be drawn to a new product in the first place, other measures are needed and, hence, different data (strictly speaking many authors do not even call it advertising initially, they are more likely to refer to

---

[17] http://www.offremedia.com/voir-article/le-barometre-adexchange-vivaki-aod-pour-100matic-/reload-vivaki-848/.
[18] http://www.iab.net/media/file/IABDigitalSimplifiedUnderstandingOnlineTrafficFraud.pdf.
[19] http://www.iab.net/viewability.

sales promotional measures[20]). The closer the customer to the right-hand side of the "purchase funnel", the more relevant becomes this so-called intent data. This refers to data points derived from the customer's specific intent to purchase the product, for instance by visiting a price comparison site, booking test drives or configuring a desired car using an online configuration tool.

On the other hand, if the objective is to introduce a new electric car onto the market for instance, the purchase interest must be created in the first place. Frequently the first step needs to identify the appropriate target group or at least fine-tune it. In these cases, "interest data" expressing more general customer preferences is of great significance. Psychometric and segmentation processes can also be used, as well as data-driven approaches that incorporate statistical processes. For instance, it is possible to determine a potential score for other customers based on the behavior data of people who purchased a certain vehicle. In other words, an algorithm extrapolates particular behavioral patterns in the early phase of confrontation with a product and uses them to identify potential customers.[21]

3.2 The "Harder" the Better

Available data is often evaluated against other data according to how "hard" it is. For instance, visiting a comparison site for electricity rates is seen as "harder" than a general interest in alternative energy, which could for instance be deduced from the viewing of certain content. In addition, it is frequently the case that data that can be generated from specific "close to purchase" events is also more suitable for the optimisation of advertising.

Nevertheless, some degree of caution is advisable, in particular for two reasons: For one thing, data often is already relatively out-dated by the time it is available online: it is immediately obvious that a customer who has just bought a car is probably the worst possible data point for further car advertising. In this situation a computed data point for other potentially interested parties can surprisingly work better, even if the data is meant to be used for sales-oriented measures. Secondly, the attribution of certain data points to products and purchase intent is often relatively complicated and prone to errors.[22] For instance, visitors of a website providing financial information might be less receptive to an investment fund product than a wellness-oriented target group with a higher income. In many cases empirical investigation (i.e. using A/B tests and involving diverse data sources) needs to be done initially in order to work out which data points work for a particular campaign goal, whereas an ex-ante categorisation in "hard" vs. "soft" data is less promising.

---

[20] Kotler, Philip & Bliemel, Friedhelm. (2001). Marketing-Management, p. 914.

[21] Wikipedia: Predictive Behavioral Targeting. URL: http://de.wikipedia.org/wiki/Predictive_Behavioral_Targeting.

[22] Particularly the media planner in day-to-day observation who is presumably subject to further evaluation errors such as attribution errors or assuming apparent correlation, cf. Kardes, Frank R. (2001). Consumer Behavior and Managerial Decision Making, p. 13 & 97.

3.3 Data Reach as a Critical Criterion

A special requirement imposed on data sources for online advertising is a result of the way in which branding campaigns are planned and what their objectives are. An advertising customer with a branding campaign typically wants to reach a certain proportion of the potential target group within a specific time frame with a minimum exposure frequency, for example, expose the ad to two million men with above-average income and an interest in cars within 4 weeks for at least three times.[23]

Many impact models, and in particular ad customer's sales expectations and plans are dependent on achieving these technical criteria. For this reason, "unspent advertising budgets" are not a good option for the people involved, not just because of the operational overheads resulting from reversals/credits etc. Thus, it must be feasible to achieve the targeted reach and exposure frequency as well. The reach requirement is far more essential to branding campaigns than for response-oriented sales advertising.

On top of that there is a second problem that usually makes things more difficult with regard to "data reach": brand advertising usually serves to attract attention first and foremost of potential customers to a product or brand. For this reason, the probability of these people having already been exposed to the product is relatively low – they have often not even had contact with the product category or related topics. After all, the point of advertising to invoke needs and to make things seem necessary until the consumer shows an interest in the product.

Both of these factors combined create a problem when it comes to delivery of data-driven online advertising, which is difficult to resolve. There is not usually enough data available – and: data sources available rarely have sufficient reach to support an entire campaign on their own.

So if an advertiser wants to introduce a new hair care product and a data provider offers him 100,000 data points of people who belong to the right target group, then chances are relatively high that this data source will not be used. This is because the customer wants to target ten times the number of people, and a mixture of data sources (or a mixed batch with and without data) leads to serious resulting problems when the campaign is being optimised and evaluated.

For these reasons the net reach of available data sources, that is the percentage of the targeted group that can presumably be covered with a data source, next to criteria concerning the origin and quality of the data, is a key to planning a campaign like this.

4 Data Protection in the Programmatic Advertising Environment

The data protection requirements for online advertising are too complex to be described adequately in one single paragraph.

---

[23] cf. Kotler, Philip & Bliemel, Friedhelm. (2001). Marketing-Management, p. 910.

Furthermore, numerous reforms to the current national regulation (e.g. the German Telemedia Act) are anticipated with the planned EU General Data Protection Regulation[24] once it comes into force.

For that reason, this paper only addresses a few specific aspects associated with Programmatic Advertising.

Despite the complexity of the detail, current data protection principles in many countries actually take a relatively straightforward and understandable approach. The internet user visits a website and should be able to rely on the fact that the site handles his/her data in a responsible manner. Legal Frameworks actually refer to the "responsible authority": the point of contact for complaints, and information requests etc.[25]

As the internet advertising ecosystem has become increasingly differentiated over the past few years, many websites work together as a group. This means that other companies carry out their marketing, and they are also cooperating with an increasing number of service providers and technology platforms. To preserve legal control yet retain the principle detailed above there is "commissioned data processing". This is a legal device that permits a third party to be given access to a website's usage data without it being transferred to a third party in legal terms. The recipient of commissioned data processing works under strict conditions on behalf of the website concerned.

This device is admittedly relatively complicated, but it has been possible to use it frequently in the past to regulate even quite complex situations in such a way that it was still possible to identify clear accountability from the perspective of the user (and in terms of data protection).

With Programmatic Advertising it is becoming increasingly difficult to adhere to the above principles. Programmatic advertising is dependent on large-scale mutual data exchange between partners and technologies, including what is known as "ID-Syncing", in other words technologies attempt to synchronise their data using particular user identification characteristics – often in a bulk manner even if no direct ad-exposure took place. Otherwise shared data would be of relatively minimal value if the purpose could not be to expose a specific user to more relevant advertising at some point.

Two things are happening during this data synchronisation, which are highly problematic for data protection: firstly, huge amounts of stored data from a variety of sources can potentially be generated, which could be made available in its full range of possible combinations. The risk of data misuse or even simply the use of the collected data in a way not anticipated by the user increased dramatically in Programmatic Advertising because of this.

---

[24] Albrecht, Jan Philipp: Alles wichtige zur Datenschutzreform (Everything you need to know about data protection regulation). URL: http://www.janalbrecht.eu/themen/datenschutz-und-netzpolitik/alles-wichtige-zur-datenschutzreform.html.

[25] Kommentare und Erläuterungen des BFDI zum BDSG (Comments and explanations from the Federal Commissioner for Data Protection and Freedom of Information on the Federal Data Protection Act). URL: http://www.bfdi.bund.de/bfdi_wiki/index.php/3_BDSG_Kommentar_Absatz_7.

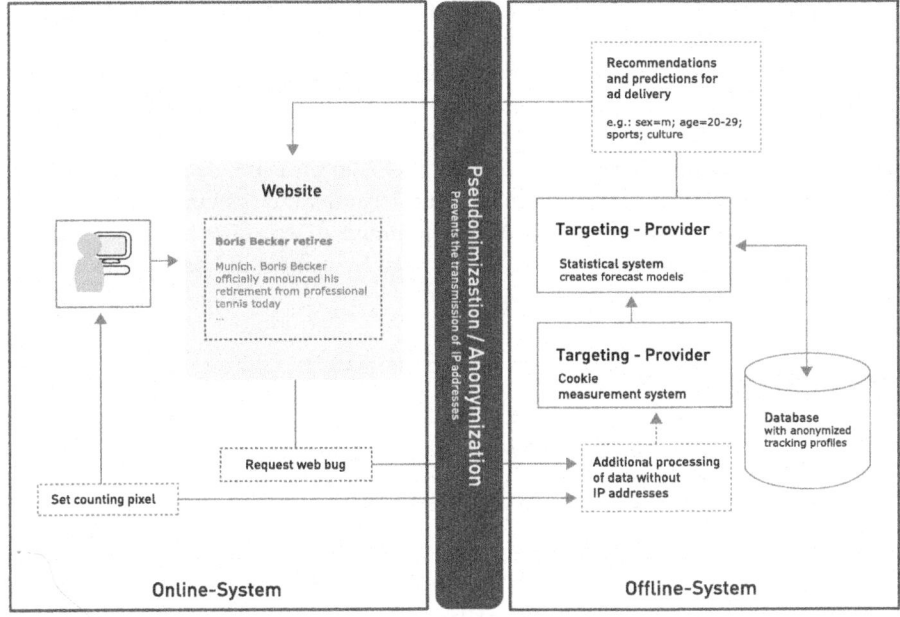

Fig. 2 Pseudonymisation process (Source: nugg.ad, 2014)

To counter this risk, investment in data protection technologies and the appropriate precautions and protective measures, as well as transparency tools, are needed urgently. For instance, it is conceivable that special pseudonymisation measures could be implemented for data synchronisation between different Programmatic Advertising players, so that it is not the actual ID of a user that is transferred from system A to system B each time, but a non-traceable derivation of that user (known as "hashing",[26] Fig. 2).

Furthermore, measures must be in place to make the process as transparent and understandable as possible for the user, for instance by using appropriate online tools that show which data pertaining to the user has been shared with which other platforms. A reliable propagation of delete requests from a particular user across all platforms involved in this data exchange would also be a key requirement in order to strengthen confidence in Programmatic Advertising platforms and to achieve a similar level of data protection as provided by existing technologies.

5 Conclusion: The Ideal Set-up to Lead Programmatic Premium to Success

A significant proportion of the advertising effect of branding campaigns will not be reflected in directly measurable events such as click and conversion rate. Instead, good branding campaigns will target the mind and sometimes the heart of the

---

[26] Wikipedia: Hash function. URL: http://de.wikipedia.org/wiki/Hashfunktion.

consumer. This means that advertising has an indirect influence on purchase behavior or the affinity for a certain brand.

So it is particularly important for providers (website operators and publishers) to ensure that the requirements discussed above are fulfilled so that Programmatic Advertising can be a success for branding campaigns. Programmatic advertising systems that have the ability to use branding KPIs to measure, deliver and optimise online campaigns must be implemented. Furthermore, the systems must be able to deliver campaigns that ensure a specified minimum ad exposure frequency, as well as reaching target audiences that can be defined by using valid data points that are available for the entire reach of a campaign.

Bibliography

Felser, G. (2001). *Werbe- und Konsumentenpsychologie* [Advertising and consumer psychology] (3rd ed.). Heidelberg: Spektrum.
Kardes, F. R. (2001). *Consumer behavior and managerial decision making* (2nd ed.). Cincinnati: Pearson.
Kotler, P., & Bliemel, F. (2001). *Marketing-management* (10th ed.). Stuttgart: Schäffer-Poeschel.
von Rosenstiel, L. & Kirsch, A. (1996). *Psychologie der Werbung* [Psychology of advertising]. New edition 1996. Rosenheim: Komar-Verlag.
Schweiger, G., & Schrattenecker, G. (2009). *Werbung – Eine Einführung* [Advertising – An introduction] (6th ed.). Stuttgart: Lucius & Lucius.

Stephan Noller born in 1970, is a psychology graduate and founder of nugg.ad, a targeting technology company based in Berlin. Previously he worked at TNS Infratest, where he developed a new kind of coverage currency for the AGOF industry association to identify the online coverage of German internet media.

Fabien Magalon has been the Managing Director of LA PLACE MEDIA, a leading premium publisher sales house based in Paris. Previously he worked at Rubicon Project, where he launched the French operations. Fabien has 15 years of experience in online advertising. Just recently he moved on to become Head of Publisher Development Southern Europe for LiveRail, the SSP in Facebook's family of apps and services.

The Creative Challenge

How to Transform Programmatic Media to Dynamic Brand Messaging

Sven Weisbrich and Caroline Owens

Building brands in the digital age is somewhat different to the "old school" Marketing approach. Nowadays, we already have a diverse and ever-increasing amount of information available about consumers. At the same time, we are constantly expanding the possibilities for targeted, programmatic ad space management. These developments not only have an impact on advertising and brand building, but also change the way the entire industry interacts.

Marketeers, media agencies and media owners need to work with each other closely to intelligently take advantage of the opportunities for branding created by Programmatic Advertising. This means interdisciplinary concepts need to be developed as link for the different disciplines. A modern advertising strategy will not plan ads as disposable products for flights (An ad flight is a fixed period of advertising activity during an overall annual campaign. With a limited budget, it makes no sense to spend away advertising money in dribs and drabs over the course of a year; you are better off going all out and investing in shorter time intervals. http://www.axelspringermediapilot.de/medialexikon/index.html?action=suche&s_text=flight), but will make optimum use of the constantly expanding information available about consumer behaviour and the tracking of ad impressions. Programmatic Advertising is not only a tool for media buying efficiency – it also leads to creative solutions by applying it to brand messaging.

The upshot of all of this is that a new and exciting challenge has presented itself to the advertising industry – and those who fail to embrace it will quickly be left behind. That challenge is simple: to combine marketing automation and programmatic technologies with creative execution and creating strong brand messages.

S. Weisbrich (✉)
UM – Universal McCann GmbH, Speicherstrasse 57-59, 60327 Frankfurt, Germany
e-mail: sven.weisbrich@umww.com

C. Owens
IPG Mediabrands, 42 St Johns Square, London EC1M 4EA, UK
e-mail: caroline.owens@map-global.com

This, in turn, will lead to the following four areas of transformation:

1. Structure: Increased interaction between marketeers, agencies, ad development and media owners, which will also lead to structural changes at agencies and marketing departments. The distinction between media agencies and creative agencies will fade away.
2. Creative development: Implementation capable of stimulating a targeted reaction to consumer behaviour in real time while keeping content, ethics and technological opportunities in balance.
3. Technology: Technologies capable of implementing data and insights in ad creation and marketing, such as rule-based dynamisation of advertising content.
4. Performance: Business intelligence solutions that can efficiently and reliably optimise programmatic campaigns as well as brand messaging.

Not even the new possibilities of Programmatic Advertising remove the challenge companies face to maintain and successfully develop their brands. The extent to which we move away from traditional campaign models towards intelligent brand management programs still depends on the brand and the industry. In e-commerce companies, for example, the focus today is more on ROI than on data-driven flighting campaigns - which resonate in an increased orientation towards control via directly measurable KPIs.

What we can say for certain, though, is that today, more than ever, the communication industry is being called on to respond to technological possibilities and to translate data-based insights on consumer behaviour into ambitious and expressive creative concepts. The use of data for the dynamic, creative design of ad contents opens a whole new chapter in digital communication – a real-time brand communication.

1 Modern Programmatic Advertising

Three years ago, programmatic media buying was a rising star with a promise to shake the advertising industry out of its complacency. Today, while programmatic itself has various connotations attached to it, it has undeniably put the very change it promised into effect. While programmatic spending as a percentage of total digital investment varies anywhere between 10 % and 40 % (thereby bringing the first real challenge to Google's search based buying model), its impact is being felt across multiple channels and across the media trading ecosystem.

The ad industry in general has been very slow in catching up with technology trends historically. But the rise of programmatic technologies has forced media agencies and media owners to appreciate its role and the contribution it makes. At the heart of the change lies the desire of any marketeer: I want to pay more to talk to people who might actually buy my product as opposed to people who will not, at a time and place that is conducive to both my consumer and my brand. Even TV, for all the giant dollar spends behind it, has never fully satisfied that desire.

Programmatic has made that desire less a fantasy and more a potential reality in the near future.

Too often, the obsession is with models, and objections tend to focus around the possibility or impossibility of execution: TV stations would never agree to sell programmatically; publishers will not allow their inventory to be priced cheaply, so on and so forth. But these questions and arguments miss the point.

If we were to strip the jargon and link back to the desire of an advertiser, all that programmatic does is enable you to reach people you think are more likely to convert (used broadly), using data and technology and pay commensurately for the media. The two components to this are technology and trading. Technology allows data to be collated, behaviour to be watched in real-time, the data stream to be put together and the message delivery streamlined in order to improve effectiveness. Trading objectives are always designed to improve value and drop costs; i.e. improve effectiveness. Whether that gets done through an auction-based exchange or a private marketplace with a floor price does not deplete its value for an advertiser with the desire referenced above.

Hence the discussion around automated data-driven marketing is timely and needed. Because the real impact of the last 3 years will be felt through automation and not through programmatic; at least not the way we understand it today. Marketing Automation is the broader trend – using software and technology for all areas of Marketing, especially CRM and web-based Dialogue Marketing. Programmatic advertising is only one part of this trend: buying, placement and optimisation of media inventory based on automated processes. The logic of Marketing Automation will influence our idea of Marketing in all areas. Because whether programmatic media buying reaches the expected levels or not, the principles it has set in place will drive all media planning and media buying in the future.

As the ecosystem matures and consumers become more aware of the value of their data and seek to actively avoid irrelevant advertising, they will seek a greater value exchange in return for their information. This would enable advertisers to improve effectiveness efficiently on a much higher scale than today. Closed loop measurement would help brand affinity driven campaigns to apply the same principles that performance advertisers use in order to drive their communication messaging. This would almost make concepts such as viewability and fraud irrelevant. When you know precisely what each exposure does for your business, it's highly unlikely that you would be spending money on impressions that are not real.

That's the promise at least. And one which if we look at the history of programmatic technology will get translated into reality to a large extent. Advertisers, whose internal systems and organisations are not prepared for this approach to marketing, will falter and lose their competitive advantage.

The assumption that big budgets and clout alone will get you the value and business you desire will be consigned to history. The more nimble you are, with the ability to mix and match between data, technology and media, the more value you will get from the wider ecosystem.

It's a simple choice really and there is a massive reward for getting it right.

The pioneers of modern advertising management such as retargeting or Programmatic Advertising are primarily e-commerce companies. However, even in their advertising campaigns, retargeting and Programmatic Advertising have rarely had any impact on ad creation so far. Overlays of product illustrations are still frequently seen in static display ads, and at best, so-called dynamic pre-rolls appear with alternating product images, implemented by region. Traditional performance strategies are often only interpreted and graphically adapted online. The intelligent technological possibilities of Programmatic Advertising are not being exploited by ad creators or marketeers.

Ad creators are faced with completely new challenges, as dynamization and the real-time abilities of ads have introduced a new era in advertising. Of course, the challenge of conveying brand messages and constructing an image remains, but use of the technical possibilities will determine the situational consumer control in future. In the e-commerce industry, the quality of the creative implementation will be responsible for a significant share of a campaign's success. Or for its failure.

2   Campaigns Belong to the Digital 80s

Traditional campaign-oriented thinking is being replaced by Programmatic Advertising. By making intelligent use of creative, intelligent communication tools, the effectiveness of advertising can be controlled and measured in real-time. Similar to search engine marketing (SEM), Programmatic Advertising strategies are flexible and guided solely by key performance indicators (KPIs). The measures are not determined by a fixed campaign, but rather by controlled and measurable success of the advertising. Modern Programmatic Advertising strategies thus correspond more to a dynamic communication program than to a traditional, rigid campaign.

As advertising in the digital world becomes more dynamic, so the approach of advertisers, their creative and media service providers need to do as well. In future, the interdependency of customer relationship management (CRM), Programmatic Advertising and ad creation will be the key to an advertising company's success.

3   Strategy, Ad Creation and Media Are All Linked Together

Today, ideas, design and media planning all merge into one overarching strategy, closing up the gap in perspective between the disciplines. The challenge will be to plan ads not as disposable products for campaigns, but to take advantage of the constantly expanding information available about consumer behaviour and ad impressions. As a result, the marketing industry must face four key challenges:

1. Structural challenges for the organisation of agencies and the marketing departments of advertisers;
2. Creative challenges, such as targeted response to consumer behaviour in real time;
3. Technical challenges, such as rule-based dynamisation of advertising content;

4. Challenges that arise due to the industry's pledge to efficiently and reliably optimise programmatic campaigns.

3.1 Structures

Soon, marketing will work in different, unfamiliar ways. Changing something in its traditional structures takes a change of thinking or even culture by companies as well as agencies. At a higher level, digital advertising technology leads to an increasing fusion of sales and marketing and, at a lower level, to the fusion of the marketing sectors of strategy, planning, creation and media. The companies and agencies that will get ahead of the market will be the ones that manage to successfully implement the necessary integration processes of these disciplines.

For media agencies the dialogue between media owners and ad creators is more important than ever now that the focus is shifting further towards CRM, dialogue marketing and the technical possibilities. To this end, creative agencies need to gain a greater understanding of the options created by dynamic advertising possibilities and expand their respective expertise. For the traditional separation between media agencies and creative agencies, this means developing into integrated agency models, since in the future agencies' strengths will lie in the interconnection and dynamisation of their own offerings. Both disciplines will build up skills in the other's field whether as a joint process or individually.

For companies, there is always a tendency to put emphasis on the brand and to base online advertising on existing offline communication. This is especially true since not all advertisers are yet digitally experienced enough that they would find it easy to reposition their brand message as a technology-based ad creation. As always with new developments, the advice of advertising and media agencies is called for here to explain the possibilities and advantages of Programmatic Advertising to the client.

3.2 Ad Creation

Ad creation is not defined by the technical possibilities, but still by the brand message and the client's brand essence. But the conception of advertising strategies will in future require the skills of digital experts as well as those of traditional art directors and content developers. Technical expertise and strategic brand and concept understanding will need to be merged. Ad creators will need to learn how to better integrate the technical possibilities and dialog/CRM marketing into their concepts, without losing sight of the traditional KPIs. By using available data and dynamised ads, ad efficiency can be markedly increased.

The first step is integrating the creative aspect into the planning process upstream. Before that can happen, creative agencies absolutely need to develop a greater understanding of media and the potential of dynamic advertising possibilities. Close dialogue with and advertising agency employee training by technology providers have a positive impact on the creative result.

3.3 Technology

The technological aspect will play an important role, but will not be critical when it comes to dynamic messaging. That is why it is important that the technological side gets a better understanding of the requirements and wishes of the ad creators. Only then will it be possible to design dynamic ads that optimally reach their intended goals.

Creativity is more important than ever, and technology shouldn't get stuck by restrictions. Dynamising ads pose challenges for all involved, but can create significant added value for all sectors. Thus it is important that creative briefings contain a strong technological component.

Even strong media owners need to develop themselves and their products, both in terms of technology and content, in order to facilitate successful creative concepts in Programmatic Advertising. Many of them still focus on transformational threats instead of a desire to create a superior programmatic offering that blows away competition. However, some digital media owners are already prepared to do this and contribute considerably to the change process with their own ideas and concepts.

3.4 Business Intelligence

Today, the decisive stage of advertising implementation only begins once the wider framework for the creative implementation has been created. The constant customisation of dynamic ads to consumer behaviour and user interaction requires companies to invest significantly more time and money in the business intelligence sector. Today there are considerably more parameters that need to be integrated into an advertising strategy with a stronger technological balance than in the past. Business intelligence is the interface for performance-based control.

4 Established Practice

A few examples in practice already prove that the steps we have outlined lead to success. For example, "Wine in Black", an e-commerce company from Berlin, pursues both traditional and modern strategies: The reach-oriented control of the classic online banner and the use of ads with appealing design and individual business intelligence-controlled programmatic elements.

The intended cost per order (CPO) is initially achieved by the reach-oriented banners. However, ads with dynamic content in more complex creative designs are catching up with increasing reach. It pays to use the available customer data in dynamic ads with a higher creative input. In the event of an increasing or constant reach, the target KPIs will be better achieved with this strategy than with traditional reach-oriented ads. Additionally, the dynamic ad can be the start of a targeted journey as landing page contents, also dynamically correspond to the ad contents.

These results are confirmed by another practical example taken from the travel industry is the campaign of South African Tourism. Where Programmatic Advertising is controlled based on individual customer data and user interests, the result is a significantly improved CPO compared to traditional ads. In this case, the creative and dynamic programmatic ad achieved a 71 % lower CPO.

5 Conclusion

Today, more than ever, the advertising industry is in charge of responding to technological possibilities and to translate data-based insights on consumer behaviour into ambitious and expressive creative concepts. This not only means that ad creators are faced with new challenges; it also indicates that marketeers, agencies and media owners need to interconnect and collaborate in new, unfamiliar ways. Advertising strategies need to be geared towards the possibilities of Programmatic Advertising from the outset if these are to be used effectively. It is an exciting journey for many companies and agencies – one that will reap the rewards in the form of positive ROI developments.

In the near future, we will see a significant increase in creative concepts that make the shift from simple product illustrations to complex advertising strategies driven by business intelligence. At the same time, the transformation process in the advertising industry will continue to march onwards, contributing to a significantly increased amount of digital spending earmarked in overall advertising budgets.

The use of Programmatic Advertising is always recommended for clients with appropriate campaign objectives, not only transaction focused brands. Ad creation will not automatically change because of Programmatic Advertising. However, in the future we will start seeing a large part of campaigns not only experimenting with overlaying the shopping cart, but also actually accompanying and supporting consumers on their customer journeys. This is where technology, content, media and data come together.

Bibliography

Fulst, C. (2014). Axel Springer mediapilot. Axel Springer Media Impact GmbH & Co. KG. http://www.axelspringer-mediapilot.de, publication date: n.s., fetch date: 28 Mar 2014.
Gross, A. (2014). Gabler Wirtschaftslexikon. Springer Gabler | Springer Fachmedien Wiesbaden GmbH. http://wirtschaftslexikon.gabler.de/Definition/key-performance-indicator-kpi.html, publication date: n.s., fetch date: 28 Mar 2014.
Mattscheck, M. (2014). Onlinemarkting Praxis. Onlinemarketing-Praxis. http://www.onlinemarketing-praxis.de, publication date: n.s., fetch date: 28 Mar 2014.
Promny, T. (2014). Online Marketing.de. Velvet Ventures GmbH. http://onlinemarketing.de, publication date: n.s., fetch date: 28 Mar 2014.
Szilagyi, B. (2014). Personology. Personology GmbH. http://www.personology.de, publication date: n.s., fetch date: 28 Mar 2014.

Sven Weisbrich In February 2013 Sven joined UM Germany (IPG Mediabrands) as Chief Executive Officer. He started his career as media planner at Mindshare in 1998 before he focused on digital strategies as UM's Director Digital. In 2008 he switched industries and moved over to the creative side, heading up strategic planning at Razorfish (Publicis Groupe) to lead new business initiatives and invent Connections Planning as their new discipline. In 20011 he took over the digital business at Wunderman Germany (WPP) as Chief Digital Officer working closely with global brands such as Microsoft, Lufthansa and Vodafone.

Today, Sven's role as UM's CEO is to implement the global vision of IPG Mediabrands. Within the last year he hired new talents from other industries (marketers, publishers, science organizations) and worked with the UM leadership team to craft the transformation strategy for UM. As a result, the agency service portfolio was diversified by launching the disciplines such as UM Marketing Intelligence (research and data science) and UM Studios (content, brand experience), to give brands a competitive advantage.

Since February 2015 Sven is also in charge of the network's digital product in Germany, MAP (Mediabrands Audience Platform), as Managing Director. Sven's mission is to grow the digital business of this agency. Further he plans to combine topics like performance, content and mobile in a smarter way with more intelligent, creative and data-driven benefits for IPG Mediabrands' clients.

Sven regularly speaks at conferences, works as lecturer at business and marketing universities and jury member at international award shows such as Eurobest and New Media Award.

Caroline Owens is General Manager for IPG Mediabrands' digital portfolio across key markets globally. She has been at the forefront of digital for over 10 years, having worked at two of the top five networks – WPP and IPG. During this time she has witnessed unprecedented change in the digital and technological landscape, and has been instrumental in educating a wide range of brands around how to embrace this exciting new world.

Caroline started out in paid search – where bid management and automation has always been central -and over time evolved to become a full-service digital expert. Most recently with IPG Mediabrands, she was a key driver in educating clients around the launch of programmatic, working with brands each step of the way to embrace the new opportunities it brought.

Her current role as a General Manager for Mediabrands Audience Platform sees her supporting digital development and growth in the media network's 14 key markets outside of the US. This involves ensuring the group's digital product and proposition evolves at pace (the integration of Data Services is a recent example of this) as well as seeing to it that international markets are using best-in-class technology to deliver efficiencies in management and effectiveness in campaign performance.

Unleashing the Power of Greater Creatives for Brands

Chip Meyers and Christian Muche

Standards should be dead; standards have held back the digital industry for years. Finally after almost two decades the leaderboard and skyscraper are finally going to the grave – there's no money in them; the brands don't want to use them; and publishers are getting rid of them left and right.

1   High Impact Ads in the Programmatic Market Place

Programmatic Advertising has been limited by its reliance on standard units, even ones like the IAB's Rising Stars. Complex, high-impact units (think rich media and video) tend to fall under the purview of direct sales because they require custom integrations and seemingly endless rounds of testing. Constant hand-holding equals a lack of automation, depriving elaborate branded executions the glory of scale. There are major massive offline budgets that have not yet even hit the digital space as there is a creative problem in Programmatic Advertising.

And until the creative problem is solved, Programmatic Advertising – and really, Real-Time Bidding (RTB) – won't truly take off.

There is a revolution afoot to bringing non-standard custom and rich media units (Fig. 1) into RTB streams. Typically if an agency has some crazy custom creative for a campaign, it will send out RFPs (Request for Proposal) to maybe the top 50 Comscore sites. After some of these pubs give the ok, the conversations start between development, tech and other departments regarding integration and

---

C. Meyers (✉)
Reactx, 3015 Main Street, suite 480, 90405 Santa Monica, CA, USA
e-mail: chip@reactx.com

C. Muche
dmexco, Messeplatz 1, 50679 Cologne, Germany
e-mail: c.muche@kdme.de

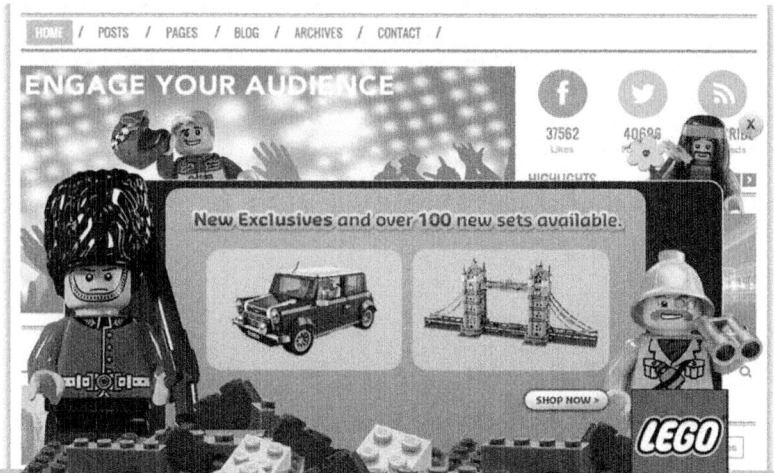

Fig. 1 Non-standard advertising (own image)

Fig. 2 Cross-device advertising (own image)

execution. But technology is enabling that all to be bypassed and instead target audiences across thousands of sites via RTB to receive your high impact creative.

Technology exists to determine browser type, screen size, presence of iframes, kind of device and more, basically certifying publishers and placements in real-time. Once understanding how a placement will look to a user in real-time, the technology can determine how an ad can be displayed properly within milliseconds. It works for responsive sites and HTML5 creatives that enable a single creative to be appropriately displayed across desktop, mobile and tablet (Fig. 2). And view ability is actually baked in to the buys.

Unleashing the Power of Greater Creatives for Brands 133

Fig. 3 High-impact video ads (own images)

Basically a platform-side integration with this type of technology removes the need for manual verification of creative by publishers. An audit tool allows buyers to check their creative before entering the exchange. Then by piggybacking tags remotely, the technology gets piggybacked onto existing ad server code and site placements. No need to place tags on sites any longer.

On the Programmatic Advertising front, talk of programmatic direct and private exchanges has moved the wild west of RTB out of the headlines. Well, not entirely – open RTB has mainly been in the news because of the bot menace and widespread questionable traffic on the exchanges.

So What's it going to take to make open RTB really flourish as intended? Creative seems a good bet.

Programmatic media is the norm; creative production needs to keep up. While automated transactional technology has come a long way, nothing much has changed on the creative front. Now's the time for innovation, particularly since advertisers are using digital video (Fig. 3) following the introduction of panel-based metrics to the channel. Video is simply merely a component of high-impact ads.

Programmatic Advertising has two central tenets: transactional efficiency and targeting audiences across the wealth of the Internet. It's arguable that open RTB has been stuck in the direct-response rut because creative options have been limited. High impact ads (with video that can be measured with panel-based metrics) traded through the exchanges already score impressive clearing prices, and could be the rising tide that lifts all digital boats.

First, brands pump up digital spend, devoting increased branding dollars to data targeting with custom creative. This jumps into the hands of agency trading desks, always hungry for brand dollars, working the exchanges. Publishers – particularly those beyond the top 100 sites who have high-value audiences – could see ramping revenue from unsold inventory, and increased competition in the channels may drive up all CPMs.

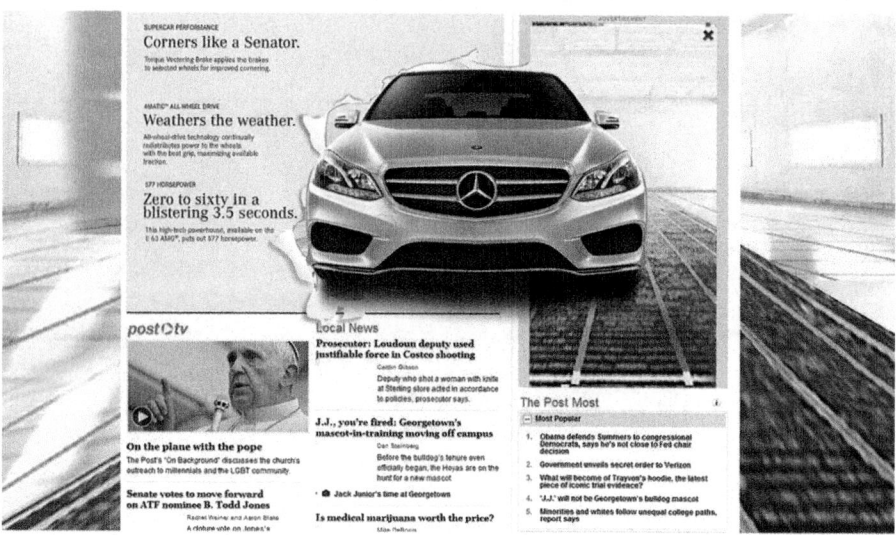

Fig. 4 Programmatic homepage take-over (own image)

This is truly the race to the top of Programmatic Advertising, rather than the race to the bottom that everyone is always complaining about. But even though the map is drawn out, we won't see the change overnight, particularly when even the network mentality lives on since many buyers using RTB still ask site lists. Still we might be witnessing the beginning of a new creative technology battle, one that will highlight the revenue-driving power of Programmatic Advertising over its efficiency benefits.

This is the huge mindset change happening. Agencies and publishers have had a standard way of doing this for years, through the RFP process and such.

Everything they have been doing for the last 10–15 years, you can now do in an automated fashion at scale but that takes time to digest.

Today, a brand's capacity to develop high-impact, custom ads is more of a creative issue than a tech issue. As issues go, it's not even a daunting one. Yes, there is more design work involved, but it is work that any professional designer can do comfortably. The design tools are the same. The creative for a takeover can be produced in a day by a design team with chops (Fig. 4).

Taking the work in-house to create high-impact ads, in conjunction with a brand's in-house Programmatic Advertising team, is inexpensive and easy, in relation to the overall scope of a digital campaign.

The benefits for bringing creative in-house for high-impact Programmatic Advertising campaigns are more than worth the effort, in terms of unified brand messaging and efficient development and deployment. If anything is going to hold up a brand from taking these tasks in-house, it shouldn't be lack of knowledge that it's even possible.

For many major brands, programmatic ad buying is no longer an experiment – it's part of their long-term marketing strategy. With brands committing themselves to the method, the next step for more and more businesses is to take the development of these campaigns in-house.

Free from reliance on agencies, in-house teams can create campaigns that act and adapt quickly and without roping in a third party. That kind of immediate action is important in such a fast-paced environment as Programmatic Advertising.

There's one place, though, where brands are lagging in taking the work in-house: developing ad campaigns specifically for the programmatic marketplace. Here, we still see brands taking extra steps. Commonly, a brand will run a programmatic campaign through a DSP, and buy direct from publishers for their high-impact campaigns. But these days, it's largely unnecessary to split efforts this way.

The DSPs are increasingly adept at deploying high-impact campaigns via Programmatic Advertising. The brands don't always know this, though. This leaves the DSPs in a position of educating their clients about their high-impact capabilities. It shouldn't be this way. High-impact programmatic campaigns are commonplace, and they should be common knowledge as well. The way the technology around Programmatic Advertising has advanced, the greatest barrier to brands developing high-impact campaigns in-house is the lack of knowledge that they can do so.

A key reason why brands haven't jumped on this train yet is that the technology to make it possible is so new – it didn't even exist a few years ago.

Brands historically hadn't thought to use Programmatic Advertising for high-impact campaigns because of programmatic standardized ad sizes.

A high-impact campaign is supposed to deliver a deep, nuanced and engaging brand experience, and that's prohibitive when you're constrained to a little box.

But these limitations are falling away rapidly. The sell side is increasingly doing away with standardized ad units, thanks in no small part to demands from the buy side for greater flexibility. You can see this many premium publishers, where the tiny boxes have disappeared, making way for ad units that support high-impact creative content (Fig. 5).

The buy side has developed the technology to deliver those high-impact ads in the programmatic marketplace. We're seeing fast-growing reach for high-impact campaigns, and these new possibilities in the marketplace are blowing up the creative side in digital.

People often think of high-impact campaigns as being complicated, but in reality, it's not very complicated at all. The heaviest lifting is over: The programmatic marketplace already exists. To deploy high-impact campaigns via Programmatic Advertising, we didn't have to build new ad servers.

We just had to layer new technology into it. High-impact campaigns were complicated because they were so manual – brands had to pre-qualify and strike deals with publishers. In Programmatic Advertising, that process is automated. The efficiencies in Programmatic Advertising allow brands to take high-impact ad development in-house, where they can invest money and energy in aligning ad creative for the best branding.

Fig. 5 Limitations in digital advertising are falling away (own image)

You don't need to consult the leading voices in the digital advertising industry to know that programmatic ad buying has altered the way forward-thinking brands find and engage new audiences at scale. Programmatic Advertising has proven to be a tremendous asset for brands to find their target audiences, wherever they happen to be on the web.

But not everyone is talking yet about how Programmatic Advertising is uniquely positioned to transform the way a campaign can deliver dynamic, engaging brand integrations for very specific luxury audiences.

Or how custom ad integrations (e.g., takeovers) and truly high-impact ads that are well suited for luxury marketers, can be scaled for RTB. The transformation is underway, and it's changing the way marketers and media buyers have traditionally thought about premium advertising.

To date digital campaigns have been hung up on ad formats: Which formats deliver the best results for which kind of campaign or audience? But the problem is, when we focus so much on standardization in ad formats, and on adhering to the specs of the same old limited formats, we get lost in the weeds and lose the path to effective, format-agnostic brand integration (Fig. 6).

Industry leaders including Neal Mohan of Google and Bob Lord of AOL Networks are saying that regardless of format, brands need to get serious about moving up the funnel, and that to do so we need to break out of the banner and address the consumer's experience in that moment. Nowhere is this truer than in the luxury space. We need to think beyond what *media* the consumer is engaging with, and imagine what's happening in the consumer's *life* at that point.

Fig. 6 Bringing brands to live online (own image)

Programmatic Advertising pulls in the data to provide a clear picture of a luxury brand's target consumer, at a particular moment wherever they are on the web.

In order to engage the audience at a point high enough in the funnel to deliver clear results for a campaign, brands must deliver them the right experience, at the right time. Old methods that rely on direct buys with luxury relevant media outlets – to reach niche affluent audiences – must be put to pasture. Luxury marketers need to target and reach their audiences on media properties that aren't necessarily thought of being "affluent."

To reach affluent audiences, brands traditionally would plan media buys with publishers that attract consumers of luxury goods and services. They would put out an RFP for each publisher they wanted to advertise with, and then negotiate the placement of a static ad – the same ad every time the page loaded. That process is neither targeted nor efficient. It involves hours of labor, and in the end, every visitor would see the same ad, whether that visitor is a CEO or a grad student doing research. It's a process that banks on a guarantee that a certain type of consumer will see the ad – a guarantee that has always been something of a fallacy.

Programmatic Advertising, on the other hand, can target affluent individuals by impression with high impact, custom ad integrations that are just as rich and dynamic as luxury goods you're selling (buy an audience, not a site). And thanks to automation, this means that instead of targeting 10 luxury publishers, you can target the same affluent audience across thousands of publishers.

Where you might have done a direct buy on a site like the Robb Report, maybe breaking news about business or finance is driving the affluents you're after to news sites – and maybe you could buy an impression for a user whose browsing behavior looks like an affluent on CNN, for a lower bid price than the Robb Report. On a day when there's a snowstorm in a particular region, you might be able to find frequent fliers on a weather site for a lower price than on a major airline's site.

2 Conclusion

Programmatic Advertising can provide even luxury marketers the type of premium, dynamic ads they covet – at scale and with relevant ad content – in a way that is integrated with the consumer's interests at that moment, across a wide variety of publishers, without all the haggling time and wasted impression that come with direct buys. The evolution of automation has brought RTB greater scale at a better value for when brands want to engage affluent consumers at just the right moment (Fig. 7). It's changing how we think of "premium," and of brand integrations aimed at the affluent consumer – for the better.

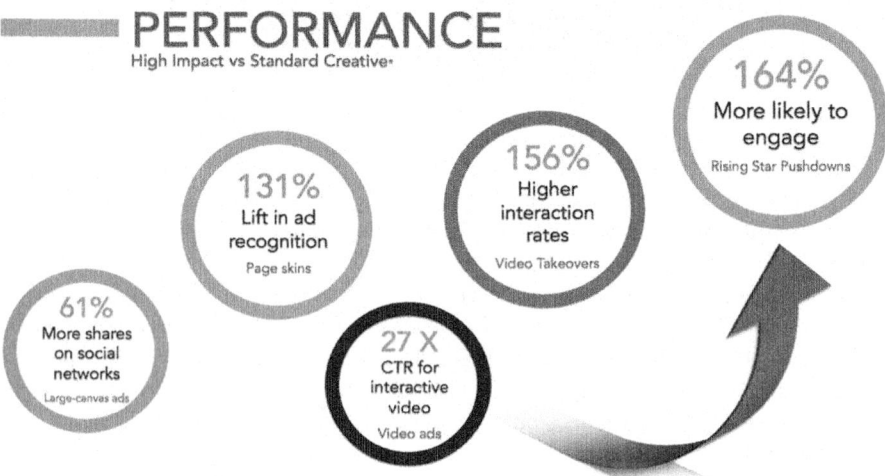

Fig. 7 Brand performance of high impact ads versus standard ad formats (own image)

Chip Meyers Founder and CEO of Venice, CA based ReactX, has over 20 years of experience in strategy, product development, and business relationships in the digital ad sector. Chip holds a B.S. degree from the University of Southern California, where his business plan in the Entrepreneur Program was deemed one of the top plans for that respective year. He has appeared on the cover of Entrepreneur Magazine as one of the top 40 Entrepreneurs under 40 and currently serves on the Advisory Board of the USC Entrepreneur Program as well as the USC Board of Counselors. He has applied and received several patents for his various inventions including the most recent application covering a radical innovation in the digital ad tech space to allow for the programmatic buying and selling of non-standard ad units across several platforms.

Christian Muche 49, is responsible for the Business Development, Strategy & International divisions on the dmexco Board of Directors and together with Frank Schneider the creator of the biggest and most important annual marketing and media show. A lifetime expert with many years of experience in the online and digital business, Muche was previously the Commercial Director and Director International Sales & Marketing at Yahoo Inc. as well as Vice President Interactive Marketing at AOL Europe. He was also the Founder and Chairman of the Circle of Online Marketers (OVK) and a member of its overall Executive Board for several years after the organization's establishment.

Cross-Channel Real-Time Response Analysis

Burkhardt Funk and Nadia Abou Nabout

Programmatic Advertising allows advertisers to bid for single advertising impressions, i.e., each time a user visits a website advertisers can decide whether they would like to bid for the opportunity to being displayed to that specific user and at what price. Programmatic Advertising, which emerged around 2009, thereby comes with a huge amount of data that can be used for decision making purposes (e.g., bidding). This article will provide an overview of the two fundamental decision making fields in Programmatic Advertising: budget allocation across the media mix and micro decision making in Programmatic Advertising ad auctions at the individual user-level. In this article, we outline state of the art modeling techniques used in both decision making areas as well as the specific challenges faced by analysts when developing models. In addition, we present common heuristics used by practitioners and potential drawbacks related to the use of heuristics vs. statistical models.

1 Evolving Media Usage

In the past, media usage typically consisted of listening to the radio while reading the newspaper over breakfast and watching a game show on TV in the evening with the family. However, media offerings and how they are used have changed tremendously in recent years. First, the number and granularity of media channels

B. Funk (✉)
Leuphana Universität Lüneburg, Scharnhorststr. 1, Lüneburg 21335, Germany
e-mail: funk@uni.leuphana.de

N.A. Nabout
WU Vienna University of Economics and Business, Welthandelsplatz 1, Vienna 1020, Austria
e-mail: nadia.abounabout@wu.ac.at

© Springer International Publishing Switzerland 2016
O. Busch (ed.), *Programmatic Advertising*, Management for Professionals,
DOI 10.1007/978-3-319-25023-6_12

have exploded over the last 20 years. At the same time, different media channels are converging, e.g., the line between broadcast and IP-based television programming is blurring. Second, simultaneous usage of multiple media offerings at the same time has become a widespread phenomenon, called "second screen". In 2014, according to a study by InMobi, more than half of viewers use smartphones or tablets while watching TV (66 % of the users in the US, around 60 % of the users in Germany and the UK).

Growing media diversity and the capability to quantify advertising response in new media offer advertisers new ways to communicate with customers and optimize communication across media channels. Traditional media planning, meaning the allocation of vast budgets to a few media channels (i.e., TV, radio, print), changed irrevocably when search engine advertising (SEA) and Programmatic Advertising moved in along with the evermore fragmented world of publishing. So now advertisers need to answer the questions: Which media channels should we invest in? How can we quantify the success of individual channels? Should a specific user see one more ads or is the number of ads shown to the user already sufficient to convert her? What kind of ad should we show the user and how much money should we bid for the opportunity to show her our ad?

2 Why Measuring Advertising Effectiveness in Programmatic Advertising?

Analyses of the effects of advertising have been conducted in scientific and practical applications for more than 50 years, with the intention of determining variables that impact advertising response and aiding in decision making regarding the allocation of budgets.

Online advertising was added to the mix about 15 years ago, with the mantra of unlimited measurability of advertising response. For more than a decade, new business models emerged and media agencies focused on so-called performance marketing measuring and analyzing impressions, clicks, and conversions. While Programmatic Advertising is still rather new and emerged around 2009, we already see the development of various management heuristics that also seem to create the illusion of complete measurability and controllability. However, it has become apparent that, even with the extensive tracking options available, it is impossible to fully measure the impact and effectiveness of advertising without uncertainty.

One of the key questions when deciding on the objective of measuring advertising effectiveness is: Which decisions should and can be made to manage ads based on their effectiveness? This question is closely linked to the two decision levels shown in Fig. 1. The upper level deals with the dynamic allocation of budgets across channels, i.e., how much of the total budget should be spent on Programmatic Advertising and when. These allocations frequently use performance indicators such as cost per mille, GRP, cost per click, and cost per order, that are rather easy to calculate. Such indicators can be compared across channels when working towards a certain goal, such as maximizing the number of new customers

Goal	Method
Budget allocation across channels	• Attribution modeling (e.g. last cookie, bathtub model) • Simple measures (e.g. CPO/CPL per channel) • Time series analysis of advertising effectiveness
Programmatic Advertising (PA) automation and optimization on a user level	• Rule based decision making • Classification methods • Markov models

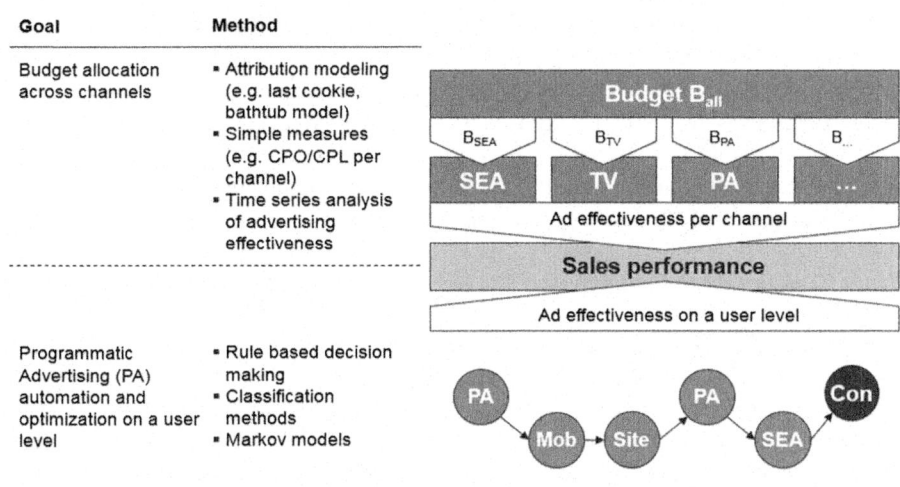

Fig. 1 Measuring advertising effectiveness and decision making levels (PA stands for Programmatic Advertising)

given a specific budget. However, they are merely supplemental aids applied to support decision making and do not necessarily lead to optimized advertising effectiveness.

Finding a statistically sound solution to optimal budget allocation is extremely difficult. It assumes that it is possible to consider the marginal costs (saturation effects) of the individual channels as they relate to the budget and to time. Budget allocation over time, also called pacing, has become an important task in Programmatic Advertising. Heise et al. (2014), for instance, develop a profit-maximizing pacing algorithm that allocates budgets taking into account a time slot's (e.g., an hour) profitability. This, however, requires that analysts are able to estimate each time slot's profitability based on clickthrough and conversion rates. But often the data as well as the skills required for decision making in that area are not available (Dinner et al. 2011). So deciding on proper budget allocation across channels is often based on experience, the performance indicators mentioned above, and on "gut feeling".

The second decision making area lies in controlling advertising impressions at the individual user-level and closely relates to so-called customer journey analysis. The fourth section of this article presents statistical models that can be applied to support decisions at the individual user-level.

In summary, there are two essential objectives of measuring advertising effectiveness: One is to allocate the available budget across channels and time (upper level of Fig. 1), and the other is to best manage advertising at the micro-level (lower level of Fig. 1).

3   Challenges and Evaluation Criteria of Methods

The development of models and methods to evaluate and manage Programmatic Advertising is associated with numerous challenges. When evaluating the different methods available to the analyst, it is crucial to understand whether and to what extent various challenges are addressed by these methods. We start by describing the challenges analysts in Programmatic Advertising are faced with:

Huge Amount of Unstructured Data For the advertisers, simply tracking ad impressions means collecting vast amounts of data on display advertising. Programmatic Advertising adds another layer, because participating in a Programmatic Advertising auction does not always lead to an ad impression. Let's assume that an advertiser wins only one in 20 auctions, it would mean that 20 times more data is generated than with traditional display advertising. Each Programmatic Advertising bid request also contains extensive data on placement. The applied methods must be able to deal with such vast quantities of data – easily 100 GB per day – and be able to process it (Stange and Funk 2014).

Few Primary, Many Secondary Attributes The data available today is characterized by large amounts of data sets, whereby each individual data set possesses only a few attributes. But because of the many manifestations of each attribute (e.g. the many publishers, peripheries, themes) and time structure of the data, a multitude of secondary attributes can be derived from this data. In order to make sense of the data, it is reasonable to categorize the various attributes: For example, instead of specifying a certain publisher, the analyst should use a context categorization (e.g. financial context instead of Wall Street online) in the evaluation of the advertising contact. The time structure of the customer journey also offers virtually unlimited ways to derive secondary attributes, such as: How active was the user in the last hour? How many of the advertiser's ads has the user seen in the last 30 days? How did she react to the ads (click/search/on-page)? The extent to which statistical models allow the advertiser to interpret their outcome is primarily a function of the secondary attributes applied. Choosing the right model requires experience, knowledge of user behavior, and an experimental, ongoing process.

Data Does Not Show the Whole Picture The data gained from Programmatic Advertising can only provide a limited view of the users' behavior. Even if there is other data, e.g. socio-demographic, available in addition to the customer journey data, a prediction of how users will behave in the future based solely on data is very uncertain: There is never a complete picture of the user, her preferences, other (offline) influencing factors and her environment. Thus, there are two requirements that suitable methods need to fulfill: First, the model needs to be able to consider data that is not directly related to the customer journey (e.g., printed ads in offline channels, weather data, competition). Second, the method needs to allow for determining the uncertainty/predictive power of the model.

Heterogeneity and Dynamics Every user reacts differently to advertising. Thus, the ideal advertising intensity is different for each user. In the same way, the number of contacts in Programmatic Advertising and other channels needed to convert a user varies. In addition, the users can be divided into a group that responds positively to advertising and one that resists it (Nottorf 2014). Amongst users who respond positively to advertising, the probability that they will react to an ad rises with each contact. In the other group, which is typically larger, the probability decreases with each contact. Methods have to take into account this heterogeneity across users in order to be able to make optimal decisions. Furthermore, user behavior is not static; it changes over time – in the long term as well as seasonally. For Programmatic Advertising, this means that the model-based prediction of a user's click and purchase probability not only has to be re-calculated with each click, the model itself might change over time.

Cause and Effect Managing advertising is not a controlled experiment in which treatments can be implemented and steered independently of one another. Rather, advertising success and managerial action are mutually dependent resulting in potential endogeneity problems. In addition, advertisers synchronize various media channels to a certain extent such as search engine ads, Programmatic Advertising, and TV leading to the so-called problem of collinearity, which in the worst case makes it impossible to answer the question of which treatment resulted in which success. Good methods identify the problem of collinearity and show the associated uncertainty around the impact of a specific medium (Note: Only specific field experiments that vary the constellation of the media channels can solve the problem; in contrast, a method can merely reveal the problem).

Methods and the models used must deal with these challenges and Anderl et al. (2013) have derived the following requirements for models trying to quantify advertising effectiveness:

- Objectivity (Model specification is transparent, calculation is data-based)
- Predictive Accuracy (Future user behavior is predicted as precisely as possible)
- Robustness (Re-calculation and slightly modified input data produce robust results)
- Interpretability (Results of analysis can be applied at the lower level to manage ads for individual users, or they can be interpreted as the impact of individual media channels to guide budget allocation)
- Versatility (New media channels, data types, and influencing factors can be taken into consideration without extensive effort when assessing the model)
- Algorithmic Efficiency (Scalability: When calculating and applying the model, computational complexity increases only moderately with the amount of data analyzed).

4 Models that Support Decision Making

4.1 Heuristics

In practice, various heuristics are used to evaluate the performance of Programmatic Advertising and other online advertising channels. Schröter et al. (2013), for instance, describe a commonly used heuristic for attributing the advertising success to individual contacts and media channels. The approach uses those customer journeys that facilitated successful advertising to calculate the contribution of individual contacts (e.g., aggregation at the level of individual media channels or publishers). In the simplest case, each contact that is part of a customer journey containing n contacts is attributed with an n-th proportion of the success. The alternative is that contacts are attributed with different amounts of the success of advertising as a factor of their position in the customer journey, e.g., contacts at the beginning and end of the customer journey are considered to have had more impact leading to the so-called bathtub model. Another model attributes greater success to the last contacts, which emphasizes the lasting effect of ads in a user's mind. If the price per contact is also considered, this model can be used to calculate costs per order (CPO) for each contact. These costs can then be used when submitting bids in Programmatic Advertising or to allocate budgets to Programmatic Advertising. In practice, this approach is closely linked to the comprehensive term "attribution modeling".

On the one hand, this type of analysis of advertising effectiveness is simple and easy to apply, which explains its widespread use in practice. On the other hand, there are some problems with this approach and it does not completely meet the requirements stated in the previous section: First of all, the success of a certain position in the customer journey can hardly be determined a priori and usually goes beyond management's gut instinct. This, however, means that it remains unclear, which of the different attribution models might be the most suitable one. Second, time aspects of the customer journey are not depicted adequately, i.e., it makes no difference whether the first contact happened yesterday or 30 days ago. Third, the impact of different types of interactions (view, click, on-site activity) cannot be determined; they can at best be considered based on hypotheses. Fourth, the analysis does not compare successful and unsuccessful customer journeys, it looks at only the former. So the essential information as to whether the success is statistically significant and the direct result of the advertising contact is lost. And fifth, offline channels and other factors not ascertained during the customer journey cannot be examined.

In summary, the approach described above can aid in the initial assessment of advertising effectiveness in various channels and sub-channels (e.g. publishers in real-time), but it is not suited to support a bidding strategy at the level of individual users. Several rules-based methods are applied in practice. The associated rules are rather diverse (Schröter et al. 2013) and will be explained using three examples: First, retargeting addresses users who are already aware of the offering of the advertiser by visiting the website or through active interaction with other advertising channels. Second, if there is socio-demographic data available (usually compiled by third-party suppliers), it can be used to address users whose profile fits the definition of the advertiser's target group (e.g. gender, age, household income). Third, a recent innovation is offered by agencies, which use Programmatic Advertising to deliver display advertising that is precisely synchronized to the broadcasting of TV ads, increasing the probability of reaching users watching TV and using a second screen at the same time. What these three bidding strategies in Programmatic Advertising have in common is that bids are submitted following specific rules. In that sense not individual users are addressed but groups with similar characteristics. This differs from the statistical models used to control bidding described in the next section.

4.2 Statistical Models

The purpose of statistical models typically is to predict user behavior, e.g., the purchase of a certain product. In general, these models are used to calculate the impact of different alternatives (e.g., different ad designs) that can be chosen to achieve a certain goal. So the probability of a user behaving in a certain way in the future is calculated based on his behavior in the past and on the potential decision making alternatives available to the advertiser – shown mathematically: *p(future behavior/previous behavior & decision making alternative)*. Then, the alternative is selected that maximizes the probability of the desired future user behavior, taking into consideration the advertiser's costs associated with the specific option.

An example: Let's assume that an advertiser has decided about a bid for a user that she knows. The advertiser knows which ads the user has already seen and how she reacted to them (e.g. visit to website, active search). Using a statistical model, the advertiser can now predict how the additional appearance of the ad in Programmatic Advertising will affect the probability that the user will purchase a certain product. The advertiser then makes a decision based on the anticipated profit margin and the cost of the ad.

There are different approaches to developing statistical models[1]; two of them will be explained briefly. The model developed by Nottorf and Funk (2013, 2014) for Programmatic Advertising is based on the previous work of Chatterjee et al. (2003). The influence of advertising on the customer journey is represented by short-term and long-term effects (Fig. 2). The short-term effects X_{actual} represent the interactions of user i with advertising within the last hour. This can be any of hundreds of different possibilities (e.g., visual ad contact within the last hour, number of clicks or searches, on-site activity, interaction term such as "User first

---

[1] Statistical models are fundamentally able to explain complicated user behavior. They can be applied to make predictions such as which product is likely to be purchased and how high sales are anticipated to be. For the sake of simplicity, we will assume that we are attempting to predict whether or not a user will become a customer.

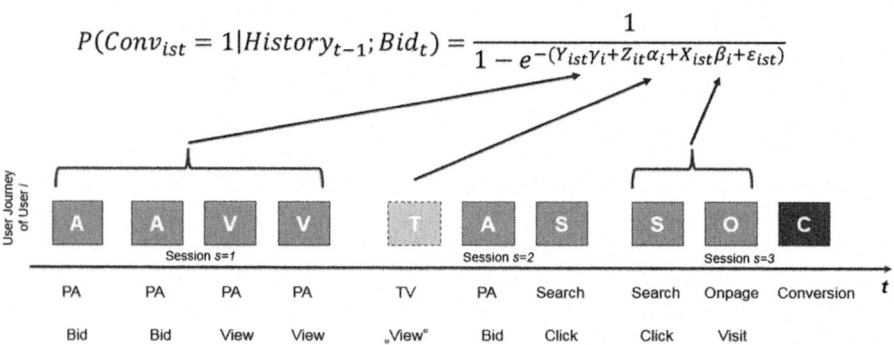

Fig. 2 Statistical model for predicting the conversion probability of a user i at a specified time t (PA again stands for Programmatic Advertising)

saw banner and then went to website via SEA"). Y_{actual} stands for advertising contacts from longer ago (e.g., last 30 days). The model (Fig. 2) can also be used to specify influencing factors not directly related to a single customer journey (Z_{actual}), e.g., a competitor's printed ads or one's own TV campaign (e.g. www. adference.de). As described in Sect. 3, specifying suitable variables is one of the essential tasks in developing the model. A case study offering a good starting point for the development of a model for a specific advertiser is that conducted by Nottorf (2014). This model is special in that it allows for differences in the way users react to advertising (heterogeneity, see above). On the one hand, this offers greater flexibility and more accurate predictions, but it also means that more effort is required for assessment of the respective model (determination of unknown parameters α_i, β_i, γ_i). It is not necessary to assign the weights in the customer journey as is the case in attribution modeling (Sect. 4.1); instead the data is used as the basis for estimation.

Once a model has been assessed, it can apply a "new" user's data within milliseconds and can be interpreted in the sense of a recommendation in Programmatic Advertising. In general, this method is considered a classification process. There are many statistical data mining methods – e.g., neural networks, support vector methods – that can be applied as soon as the variables have been specified. The forecasting quality of the models developed tends to have less to do with the selected method and more with the *astute definition and selection of variables* – taking into consideration all of the available information on the advertiser, the competitive environment, and customer behavior as well as the service/product advertised.

The second approach to be explained here is calculation of so-called Markov models. These are also intended to predict the probability of desired user behavior (see above). In general, Markov models describe potential transitions from a system's states and their probabilities. They can be shown as a graphic network model, with the nodes representing the states and the edges the transition probabilities. The order of a Markov model indicates how many of the states previously visited have an effect on the probability of transition to the next state, i.e., the likelihood of transitioning to the next state depends solely on the current state. A second-order Markov model takes into consideration the current and the previous state.

How can these models be applied to Programmatic Advertising? In a simple model, interaction with a media channel (e.g. Programmatic Advertising or SEA) as well as purchasing a product are both interpreted as states. The customer journey can be considered as a series of transitions between states. The transition probabilities between the states can be calculated using historical customer journey data, allowing the purchase probability at any point in time to be estimated for a customer journey that has not yet been assessed. Archak et al. (2010) suggest using first-order Markov models to model the impact of decisions related to online advertising (submitting a bid in Programmatic Advertising) applying the *removal effect*. This is done by removing the node in question, in this case Programmatic Advertising, from the network model and then using the remaining network to calculate how the purchase probability changes. If adding the node (placing an ad) has a positive effect and the cost is justified, a bid is submitted. Anderl et al. (2013) apply higher-order Markov models and use case studies to prove the forecasting quality of their approach.

5 Conclusion

Analysis of advertising effectiveness in real-time is still in its infancy, so it should not be seen as a project but as an ongoing process (a process requiring extensive effort).

To successfully establish continuous and action-oriented analysis of advertising effectiveness, advertisers must be *aware of the objective* and must determine the decision making level of online advertising (budget allocation across channels or placement of ads for individual users) at which the results of the analysis should be used.

The analysis of advertising effectiveness as well as its interpretation and use pose a number of challenges to advertisers. The requirements described in Sect. 3 can serve as a general guideline for selecting suitable methods and models. Caution is essential when applying and interpreting simple management heuristics and their promised cure-alls. Identifying significant control parameters in Programmatic Advertising and predicting user behavior will luckily remain uncertain and at the same time exciting for practical applications as well as for research.

Bibliography

Anderl, E., Becker, I., Wangenheim, F., & Schumann, J. (2013). Putting attribution to work: A graph-based framework for attribution modeling in managerial practice. *Social Science Research Network* 2343077.

Archak, N., Vahab, S. M., & Muthukrishnan, S. (2010). *Mining advertiser-specific user behavior using adfactors*. Proceedings of the 19th International Conference on World Wide Web 2010, 31–40
Chatterjee, P., Hoffman, D. L., & Novak, T. P. (2003). Modeling the clickstream: Implications for web-based advertising efforts. *Marketing Science, 22*(4), 520–541.
Dinner, I., van Heerde, H., & Neslin, S. (2011) Driving online and offline sales: The cross-channel effects of digital versus traditional advertising. *Social Science Research Network* 1955653.
Heise, M., Abou Nabout, N., & Skiera, B. (2014). *Profit-maximizing pacing for budget allocation over time in real-time display advertising*. Working Paper, Goethe University Frankfurt, Vienna University of Economics and Business.
Nottorf, F. (2014). Modeling the clickstream across multiple online advertising channels using a binary logit with Bayesian mixture of normal. *Electronic Commerce Research and Applications, 13*(1), 45–55.
Nottorf, F., & Funk, B. (2013). *The economic value of clickstream data from an advertiser's perspective*. Proceedings of the 21st ECIS conference, Utrecht.
Olejnik, L., Tran M.-D., & Castelluccia, C. (2014). *Selling off privacy at auction*. Proceedings of the NDSS 2014, San Diego.
Schröter, A., Westermeyer, P., Müller, C., Schlottke, T., & Wendels, C. (2013). *Real time advertising – Funktionsweise, Akteure und Strategien*, http://rtb-buch.de. Sept 2013.
Stange, M., & Funk, B. (2014). *How big big data needs to be? The learning curve in Bayesian user journey analysis*. Working Paper, Leuphana Universität.
Yuan, S., Wang, J., & Zhao, X. (2013). *Real-time bidding for online advertising: Measurement and analysis*. Proceedings of the ADKDD 2013, ArXiv, 1306.6542.

Prof. Dr. Burkhardt Funk is a professor of Business Information Technology, particularly E-business, at Leuphana University in Lüneburg, Germany. His research focuses on quantitative model building for decision making in E-commerce and online advertising. Funk has published over 50 papers and three textbooks and has conducted several third party funded projects with partners from industry and research institutions. Funk shares his knowledge, acting as a consultant to companies on the subject of creating and managing models demonstrating the impact of advertising. He continues to be actively involved in or providing consulting on founding various companies.

Prof. Dr. Nadia Abou Nabout is a professor of Interactive Marketing & Social Media at WU Wien, Austria (http://www.wu.ac.at/imsm/team/abounabout/). Her research is located at the interface of marketing and information systems and focuses on radically new technologies in marketing such as real-time bidding and programmatic advertising. In her work, she aims to help companies make better marketing decisions and builds upon extensive industry collaborations. Abou Nabout's research has been published in leading journals of the field (Marketing Science, International Journal of Research in Marketing, Journal of Retailing). Together with Bernd Skiera, she was one of three finalists in the "Gary L. Lilien 2011–12 ISMS-MSI Practice Prize Competition".

The Contribution of Measurement in a Cross-Device, Data-Driven, Real-Time Marketing World

Niko Marcel Waesche, Tilman Rotberg, and Florian Renz

Considering the advanced algorithms of programmatic advertising, its great efficiencies and its advanced capabilities to address consumers at the point of decision during their purchase journey, one could argue that third party advertising effectiveness measurement is not needed any more. Some believe that programmatic is a magic solution, that finally liberates the advertising industry from cumbersome measurement exercises.

Programmatic advertising is not the magic solution that solves all problems. The algorithms of programmatic advertising do not make measurement approaches redundant. On the contrary, machine algorithms and panel-based data complement each other, and validated measurement techniques are needed in order to improve the performance of campaigns, which are using programmatic advertising. The good news: We do not need to reinvent the wheel and create all-new measurement tools just for programmatic; instead we can repurpose and slightly modify them. Measurement offers powerful tools to address some of the known concerns with programmatic advertising and contributes to its acceptance and maturity in the industry and beyond.

It is clear that programmatic advertising represents new, unparalleled opportunities for advertisers and forward-thinking media groups, but it is also a black box, characterized by complexity, a lack of transparency and data challenges. There also is a lingering consumer trust issue linked to the use of personal data by programmatic engines which premium advertisers cannot ignore. There is an urgency to address these challenges, since the development of programmatic

---

N.M. Waesche (✉) • T. Rotberg
GfK, Nordwestring 101, 90419 Nuremberg, Germany
e-mail: niko.waesche@gfk.com; tilman.rotberg@gfk.com

F. Renz
GfK, Herrengraben 3, 20459 Hamburg, Germany
e-mail: florian.renz@gfk.com

advertising is not stopping; it is spreading from display into new areas, into video and mobile advertising and beyond.

In this chapter, the authors briefly discuss the success of programmatic and its spread to new areas, highlighting the key challenges of programmatic and how they relate to measurement. In the third part of the chapter, the authors illustrate how validated (panel-based) measurement techniques are essential in order to address some of these challenges and in order to improve the performance of programmatically driven campaigns. Finally, the chapter ends with an outlook describing a world in which consumers themselves control and manage some of the data that is used by programmatic engines. In the opinions of the authors, this development will contribute to an overall more mature programmatic advertising ecosystem.

1 Measurement Faces a New Marketing World

Programmatic advertising is about the data-driven, automated and real-time matching of consumers with ads for goods and services. In the United States, programmatic already today generates $10bn of sales and makes up 45 % of all digital display advertising sold. ["US Programmatic Ad Spend Tops $10 Billion This Year, to Double by 2016," eMarketer, October 16, 2014] The reasons for this success are the increasing attention on advertising invests, both in terms of lower costs and also higher revenues generated from ads, the rapidly increasing amount of available data from consumers, as well as the more and more sophisticated algorithms of ad engines and their growing popularity.

While programmatic was introduced in the context of display, it is now expanding far beyond that: A key driver is mobile; already today, mobile accounts for 44 % of programmatic spend. ["US Programmatic Ad Spend Tops $10 Billion This Year, to Double by 2016," eMarketer, October 16, 2014]. But programmatic has also been applied to TV advertising; Simulmedia is one example for this development.

ESPN, uses the content preferences of their audience – captured e.g. via set top box return path data – to improve the targeting of advertising on its site. Recently, ESPN launched its own proprietary Data Management Platform (DMP) to integrate its data about its audiences with ad supply systems. [Tyler Loechner and Joe Mandese, "ESPN To Launch TV DMP: Will Enable Brands To Target Audiences, Not Ratings," MediaPost, March 26, 2015.]

Programmatic and creative is merging, too. Dynamic Creative Optimization (DCO) refers to a technology which extends the matching process between ads and consumers to the creative of an advertisement. Different creatives can be served to different targeted audiences depending on interests and demographics. In a blog post, Google's Creative Platforms Evangelist Pete Crofut speaks about "predictive creative" in which DCO is taken a step further by introducing advanced analytics to optimize the creative of an ad as it runs based on performance. [Pete Crofut, "Programmatic: Merging Data and Creative," *Think With Google Newsletter*, May 2014]

Another example for the ongoing rapid development of programmatic advertising, is that the borders between CRM and marketing are being broken down, with programmatic driving this development. Berlin-based ad tech company Sociomantic, acquired by Dunnhumby, has published a case study showing how data from L'TUR's CRM systems was used to set up targeting within programmatic advertising engines. [Sonja Dreher, "Case Study: L'TUR Boosts Sales by Combining Smart Booking Data and Programmatic Tech," Sociomantic, April 1, 2015].

The startup SO1 in Berlin has created a programmatic coupon system for in-store visits. Shoppers have a loyalty card, but do not need to register their address or contact details. This makes the system very simple to use. The card helps register a database of purchased items in the stores, it acts as a "human cookie." When shoppers enter a store, they put the card in a machine which in that real time instance creates an auction among consumer goods companies interested in promoting their items to that specific, anonymous individual with that purchase history. In order to make the purchase history more meaningful for targeting, the history is imputed across retailers based on algorithms that are trained on a database of hundreds of thousands of anonymized household purchases supplied by GfK. The semblance to programmatic is clear, in fact, it is an almost direct translation of ad tech into the world of offline retail.

With time, the real time data matching approaches developed within programmatic advertising will enter more and more areas of consumer lives. In late 2013, John Battelle, former Editor of Wired and author of "The Search" published a blog post with the slightly cumbersome title: "Why the Banner Ad is Heroic, and Adtech is Our Greatest Artifact." [John Battelle, November 17, 2013] The blog post was shared over 2000 times; it was a revelation for many because it took adtech out of its narrow confines in the ad world and described it's broader impact on society.

It is easy to see how Battelle is right and how any type of consumer intentions might be matched real-time by algorithmic systems. This applies also to the most previously private realms of life. Quantitative self-devices such as fitness trackers generate a stream of health data which could be matched real-time to up-to-the-minute disease diagnostics systems at a doctor's office.

Joined to this observation, however, comes the warning that we need to be careful how these systems work. In his blog post announcing the future ubiquity of automated systems, Thomas Davenport states that, so far, "...The trouble is, that they are not working so well." [Thomas H. Davenport, "Era 4.0: The Scary Age of Automated Networks," CIO Journal/The Wall Street Journal, April 1, 2015].

2   Barriers to Further Growth of Programmatic Advertising

Despite its increasing popularity, programmatic advertising, as it is being used in marketing today, is facing some significant challenges from the perspective of premium advertisers. These are issues with transparency, questions about data quality and sourcing, understanding the true impact within a campaign and consumer trust. Validated measurement techniques can help, however, to control some of these challenges.

2.1 Lack of Transparency in the Value Chain

A large amount of players, big and small, are seeking to capture a sliver of the rapid growth of programmatic. The famous Lumascape maps show only the tip of the iceberg of this ecosystem, depicting dozens of companies that are above a certain valuation threshold. There are, in fact, many more which are not on the maps. Often, experts who are specialized on one segment of the Lumascape have very little understanding of another. "Complexity characterizes the programmatic landscape, but so does innovation." Consolidation of the programmatic space has been predicted for a long time, but new companies keep emerging, seeking a role in the ecosystem.

For the CMO as a buyer of advertising with an interest in understanding its effectiveness, it is very difficult to navigate the complex world of programmatic. Inventory trading and arbitrage by some participants in the ecosystem have been criticized because clients on the advertiser side have been left out of the transaction. Who makes what money in the value chain is not clear for some market participants. [See, for example, Joe Mandese, "There Is A Bear In Madison Avenue's Woods, And It's Kicking Back: Analyst Says Agency Rebates Are No Bull," *MediaPost*, April 13, 2015].

This lack of transparency can create a black box. A lack of transparency can lead to higher costs and a faults assessment of the ROI/ROMI of a campaign. More importantly, the contribution of programmatic advertising within a cross-media campaign cannot be assessed accurately and attribution is difficult.

2.2 Questionable Data and Sources

Programmatic advertising requires large data sets describing consumer and audience behavior in detail. One implication is that new data sources are continuously evolving and that the quest for consumer data rapidly increases, with stakeholders in the media industry striving to enhance their control over data. In order to quench the thirst for data, specialists have already for a long time been gleaning data from many different sources, providing them to the ad ecosystem.

Some practices of consumer data sourcing are under watch and policy-makers in the US are discussing regulation. [Ryan Joe, "Sen. Jay Rockefeller Puts Acxiom, Epsilon And Experian 'On Notice,'" *AdExchanger*, December 18th, 2013.] The emphasis on gathering more and more data to fine-tune the targeting mechanisms has also led to misuse, such as the collection of data from questionable sources and those that do not respect privacy preferences.

Fraud is also an issue. A *Financial Times* article created buzz in the industry, stating that a specific digital Mercedes Benz campaign was viewed by more fraud bots than actual humans. [Robert Cookson, "Mercedes online ads viewed more by fraudster robots than humans," *Financial Times*, 26.05.14] Some have criticized the article, since it applied to only a small part of the overall digital campaign, but it was still a wakeup call. Companies must be careful where they source their data for programmatic advertising.

More and more stakeholders are becoming concerned about data quality and security. In order to control marketplace and quality, digital publishers such as NewsCorp have launched their own ad exchanges, or DMPs, and innovative broadcasters such as ESPN have followed suit [Lucia Moses, "News Corp Launches Global Private Ad Exchange New spinoff embraces programmatic buying," *AdWeek*, August 21, 2013].

Not only media groups, but also advertisers and agencies increasingly focused on the sourcing and quality of their data. Panel-based data from established research agencies is sourced with full knowledge of the respondent and quality controlled. Increasingly, it is being used as reference data by programmatic advertising.

2.3 Unclear Impact of Programmatic, Especially Related to Cross-Media and Brand Advertising Campaigns

To add to these issues related to transparency and data, there is a lot of uncertainty around the true contribution of programmatic advertising as a part of comprehensive marketing campaigns. The impact of programmatic is often overstated in a campaign because a retargeted banner is much closer to the "last click" than a TV campaign, even though the television campaign could have built up important brand awareness which led to the "last click." Attribution favors the "last click". In fact, GfK ROI analyses – based on more than 150 analyzed CPG cross-media campaigns – show there is indeed a correlation between purchase act and online clicks including the "last click" (0.44), but much more important are other factors, for example the relative size of the total gross media spending to the turnover of the product (0.76) and the impact of the creative itself (0.69).

For a CMO juggling a campaign consisting of TV, mobile and internet advertising, possibly also out-of-home, print and radio, the combined contribution of the individual elements is almost impossible to assess without any additional reporting data. Incremental reach, for example, is a challenge, has a campaign reached the same people multiple times over numerous channels or have new audiences been touched by ads on single media? One key question is, what are the weak elements of a comprehensive marketing campaign – what can be left out? The true ROI/ ROMI of programmatic in the context of a holistic campaign cannot be established without understanding effect on sales from each individual media channel and jointly.

Cross-media advertising planning is becoming the industry norm but CMOs need better information to understand how to best optimize the mix between online and offline. Up to now there is no standard currency available which is able to indicate which parts of offline and online advertising are most efficient – for each campaign individually. The industry requires cross-media key performance indicators like the real gross/net return on investment for each media channel individually.

Brand advertising puts CMOs in front of another challenge. Now that programmatic is moving into video, an immediate conversion is often not the goal. But what is the contribution of a programmatically served video ad on brand awareness? This question cannot be answered by server side data capturing clicks and conversions.

In summary, CMOs of big consumer brands are not ignoring such concerns. They are placing increased attention on transparency within the programmatic value chain and on the sources of data. They are also seeking data providing insight to cross-media and cross-device reach, brand impact and which can report on the ROI/ROMI of their campaigns.

2.4 Consumer Trust and Privacy Concerns

Consumers are becoming alarmed by the use of their data for retargeting and in more and more aspects of their lives. The trust issue is not being ignored by CMOs of premium brands. Often people do not know that their data is being collected or they have accepted it unknowingly by clicking on long and cumbersome T&Cs. The interest in privacy-hardened smartphones like the Blackphone [www.blackphone.ch], shows that some users are taking the defense of their privacy into their own hands and are even prepared to pay a premium to regain control of their digital footprints. In the perspective of some, programmatic contributes to a world in which people overconsume. Some users employ plugins like AdBlock in an attempt to remove most advertising from their browsing experience. One future might entail a continued arms race as advertisers try to force themselves onto users' screens.

Some consumers feel that they want to take back their data ownership. Several new projects have emerged out of this concern, such as the "Personal Data Fortress" Cozy Cloud [http://cozy.io/en/], a French initiative, or "intent casting" technologies such as Intently, from the UK, or Zaarly, for home services, originating in San Francisco. In "intent casting" the idea of advertising is put on its head, it is people broadcasting their purchase intentions into an automated matching and/or bidding system, not the other way around. A list of interesting development work around "Customer Commons" can be found on the ProjectVRM page. [http://blogs.law.harvard.edu/vrm/2013/02/05/vrm-development-work/]. Personal Health Record and Quantified Self, for example, are examples for bottom-up initiatives aimed at re-asserting control by people of their personal data [http://quantifiedself.com/].

In summary, there are several barriers challenging the further development of programmatic advertising, especially when it comes to its use in premium brand environments and cross-media campaigns. Validated measurement approaches can address some of these barriers, as will be described in the next section. Better measurement alone cannot solve the lingering issue of consumer trust, but here several initiatives are working on placing more control over data directly into the hands of consumers. The final outlook of this chapter will describe these efforts.

3 The Power of Panel-Based Data. Or: Measuring the Success of Programmatic Advertising

In order to measure the success of programmatic advertising and in order to optimize it, advertisers can use the data provided by consumer panels, ideally on cross-media, single-source panels, provided by market research companies as a neutral third party. Governed by industry norms and associations such as ESOMAR, market research companies have long been third-party providers of trusted and robust data to advertisers and the media industry. This role is still valid in a future, a future in which programmatic advertising plays a larger and larger part and moves from purely retargeting and real time bidding into new areas such as video, mobile and beyond.

The role of market research companies and their consumer panels can be split into two parts: (a) Direct provision of robust, privacy-safe reference data to programmatic ad tech engines for calibrating programmatic engines, for improving data quality and for fine-tuning algorithms. In this way, panel-based data can address some of the data issues related to quality and sourcing faced by programmatic. (b) Use consumer panels in order to validate reach, understand incremental reach of programmatic in a cross-media and cross-device environment, contribute to insight into brand impact and provide results about campaigns as input for ROI (Return on Investment)/ROMI (Return on Marketing Investments) calculations.

While the first role of panel data is new, the second has been the raison d'être for cross-media single source panels for several years, even before programmatic advertising came into marketing big time. Panel-based data cannot directly solve transparency issues associated with the programmatic advertising value chain. However third-party data about the effectiveness of programmatic advertising, especially within cross-media and brand marketing environments, supports ROI calculations and attribution.

3.1 Panels as Reference Data to Improve Data Quality

Market research companies do not engage directly in the advertising fulfillment process. They do have high quality sets of data derived from panel respondents with whom they have agreements with. They can provide this data aggregated and anonymized for use in marketing activities, including for programmatic advertising. One such service is the calibration of programmatic targeting. As discussed, ad tech engines use Big Data sets derived from various sources and algorithms to setup, evaluate and improve their targeting activities. High quality market research panels can be used to collect the data markers left by digital advertising. These can be checked for their accuracy regarding demographics, preferences or segments. Adjustments can be made, meaning that target groups are matched with advertising in a more accurate and less unwanted way. It is important to note that these calibration activities do not involve the panel members directly. At GfK, panel data used in such way is called "Reference Data."

3.2 Increasing Transparency and Measuring Effectiveness Through Panels

At its core, cross-media metered panels measure smartphone and desktop internet use passively, without active requirements on the participants. The most important aspect of the panel is its single-source nature, meaning that smartphone and desktop data are obtained from the same individuals. Single-source contributes to a much more accurate assessment of incremental use because, at its core, data is not fused but directly obtained. All panel participants voluntarily join the panel and agree to their data being de-identified and used for marketing purposes. This also means that demographic data is accurate. In several countries, the core panel is expanded through passive TV metering and/or scan-based purchasing data. Surveys can be added as well, for example to assess brand lift or the additional contribution of print or out-of-home advertising in the marketing mix.

We discussed above the often unclear impact of programmatic advertising, especially in a cross-media environment. This is where cross-media single-source panels are needed in order to measure the reach and effectiveness of cross-media campaigns with programmatically driven elements. An example: GfK's cross-media single-source panel was used to analyze the effectiveness of a large German TV- and digital campaign for an e-commerce brand. The objective of the analysis was to understand the link between the interest raised by TV and the combined effect on digital traffic and conversions. The purpose of the campaign was to drive as much traffic as possible to the e-commerce portal. Significant multi-million sums were invested in TV, programmatic retargeting and search advertising over the course of a few weeks.

Without panel measurement data available, marketing departments often try to create cohorts of traffic flows between TV and digital by capturing server data from the exact time slot of each TV ad. In large campaigns, however, television spots often overlap. Also, timestamps from diverse digital ad types and server visits are not always available. Medium-term effects, such as visits an hour after the TV ad, are not captured at all, leading to a decrease in the calculated ROI of TV and an underestimate of the importance of the TV campaign.

By actually capturing visits to the e-commerce portal as well as media exposure digitally and on television, GfK could demonstrate the complete effects of the ad campaign. The starting point actually is a clear picture what proportion of visits to the e-commerce portal were not in any way influenced by advertising; these visitors would have come anyway. Panel data could establish that 63 % of website visits to the ecommerce site were not influenced by the media campaign. Furthermore, by itself, both television as well as digital would have disappointed in their effectiveness. 3 % of website visits were exclusively generated by the TV campaign and 12 % through all digital initiatives, including display, search and retargeting. Together, the media mix was powerful, however. 22 % of all visits were caused by television in combination with a digital ad format. There was a high proportion of overlap in TV and digital reach, therefore, and this overlap was essential for the success of the campaign. The data could also be used to compare accurately the combined effectiveness of TV with display, retargeting and search. This data was used to improve subsequent iterations of the campaign by GfK's client.

E-commerce companies care about conversions on their portal. For FMCG clients, offline sales are essential to quantify ROI/ROMI of advertising campaigns. To obtain results, the method of logistic regression is used. The base is the purchase act in the consumer panel. The analysis models every consumer decision. For every purchase occasion the analysis takes into consideration which brands were in promotion, how the price level of the brand was, how many ad contacts the purchaser had before purchase and with which media. Causality of effects can be assured by controlling socio-demographic and loyalty effects.

The result of each evaluation shows the influence on the single purchase decision (brand choice) – independently from one another and including interaction effects. The output is the additional revenue, which is generated through each advertising activity. The KPIs are the gross/net Return on Investment, the individual uplift for each media activity.

In another project, Facebook and GfK joined forces to measure the ROI of FMCG advertising of comprehensive, exemplary ad campaigns which include a significant proportion of mobile advertising. Even though smartphone behavior is collected on GfK single-source panels, it was clear that a combination of mobile ad impression data obtained from Facebook joined with panel data such as TV views and purchases would provide the most powerful data set from which to evaluate ROI.

Facebook and GfK needed to find a way to utilize login data in a de-identified way to enable comprehensive cross-device measurement on the single source panel. A data link, which uses salt-hashed user IDs and a trusted third party acting as a "dumb wall" between Facebook and GfK allows Facebook to inform GfK about ad impressions served to individual panelists without GfK learning anything about non-panelists and without Facebook learning who the members of the GfK panel are. With this data link in place, GfK is able to conduct single-source studies that compare the impact of campaigns on TV with those on mobile using a media mix evaluator methodology.

Based on this data link and marketing mix modeling process, GfK conducted a meta-analysis of seven cross-media FMCG campaigns, each including both TV advertising and advertising on Facebook. The objective was to analyze the short-term impact of the campaign elements of the different media channels (TV vs. Facebook) on off-and-online purchases of the respective FMCG products. "Short-term impact" was defined as sales as they occurred while the campaign was running and up to two weeks thereafter. Depending on the campaign and the product, TV and Facebook performed differently; Facebook had a strong short term impact especially in those situations where the unique targeting capabilities of Facebook were beneficial. However, it is important to note, that these results cannot necessarily be transferred to mid- and long-term sales effects of advertisement campaigns run on TV or Facebook. In addition, effects on brand lift were not analyzed.

It was also discussed above that the impact of programmatic for brand advertising is often not assessed. In fact, by far the biggest proportion of ad spending is related to brand campaigns, which have much longer time horizons than it is typically measured in most digital analytics systems. Now that programmatic is moving heavily into video, an immediate conversion is often not the goal. To measure the contribution of programmatically served video ads on brand awareness, Facebook teamed up with GfK again, this time to understand the brand effects the campaign of a major automotive brand in several European countries.

The approach to measure the brand effect of the Facebook campaign was straightforward, drawing on GfK's digital capabilities but also on years of measuring brand lift for television and other "offline" campaigns. On a panel, cookies were collected related to the campaign, allowing GfK to identify panel members with ad exposure and differentiate them from a control group without brand exposure. Surveys were administered on this basis, not based on recall but on actual verifiable ad exposure. Total spontaneous brand awareness was lifted in Germany and Italy by 10 % respectively 5 % among those who had seen the ad. However, the associations of the same campaign were very different in Germany and Italy. Whereas in Germany, perception focused on a "accomplished vision" in Italy the campaign was associated with "environmentally-friendly" and "familiarity." Brand effects were very different in each country, even though the campaign was identical. The brand effects of the Facebook campaign could be compared to the microsite. The results were detailed and could be used by the marketing team as insight for future campaigns.

In summary, programmatic advertising is facing various challenges, which require a close monitoring and performance assessment of programmatically driven campaigns.

Cross-media, single-source panels, provided by market research companies, are the ideal tool for this and can function both as reference data for data used in ad engines as well as tools which measure the success of programmatic advertising.

4 Conclusion

Increasingly, consumers are submitting and managing their own data actively, and they apparently love it. One very common example are dating sites, such as OkCupid, a matching site that has thrown the top-down approach of survey and dimension design overboard in favor of user-generated dimensions for matching. Christian Rudder is also publishing research of OkCupid's experience with grassroots question design for matching on their OkTrends blog and in his book: "Dataclysm: Who We Are (When We Think No One's Looking)."

Several smart startup teams are currently focusing on re-inventing the career market based on data submitted by job seekers. They are using intelligent matching technologies to find the best fit between job seekers and companies. Craft.co in London and Jobspotting in Berlin are both working with self-managed personal data, are categorizing jobs and developing the best matching algorithms. Craft.co is categorizing employers allowing them to directly provide the most up to date matching data. Jobspotting is working with 30,000 different career tags associated with skills, background and preferences. Check out these initiatives. It is easy to imagine that these types of services will one day be the standard way that professionals and job openings come together.

What has already been happening for a long time is the closer integration between media content and ecommerce. Farfetch, a London-based e-commerce startup, and Mybestbrands, based in Munich, both match content preferences, make fashion recommendations and provide leads to third-party online shops. Neither company has any inventory, nor do they engage in shipping. [Ingrid Lunden, "Fashion Marketplace Farfetch Raises $86 M Led By DST At A $1B Valuation," TechCrunch, March 4, 2015.]

A final example in which customers submit data voluntarily in order to gain insights and deals is Moneysupermarket.com. This is a UK-based web destination for people seeking solutions to saving money on services such as home insurance, gas and electricity. In 2013, Moneysupermarket built its own private trading capabilities and integrated it with third-party DSPs and DMPs. [Jessica Davies, "Moneysupermarket builds private trading desk in data monetisation push," The Drum, April 10, 2013.]

The big players in the industry are all working on data transparency and personal control of data by their users. Facebook, for example, knows that significant responsibility comes with this personal data, and Mark Zuckerberg emphasized the control Facebook members have over their data several times at the F8 conference in March 2015. Zuckerberg is extremely aware of the dangers of intrusive advertising, having seen his biggest early competitor MySpace basically commit suicide through an overindulgence on intrusive ads [see Erik Schlie, Jörg Rheinboldt, Niko Waesche, "Simply Seven, Seven Ways to Create a Sustainable Internet Business," Palgrave Macmillan, 2011, page 88.]

The Customer Commons movement and similar bottom-up initiatives are providing alternative models to the existing ad system, pioneering "Personal Data Fortresses" and "intent casting" systems. We are moving rapidly into a world in which more and more consumers themselves control and manage data that is used by programmatic algorithms. This development is very important because it is addressing the trust issues consumers have and leads to an overall more mature programmatic advertising ecosystem which will serve all sides of the ecosystem: Advertisers, media companies and also consumers.

Niko Marcel Waesche is the Global Industry Lead for Media and Entertainment at GfK and oversees GfK's global media and digital business. Niko has worked in media for more than fifteen years. He co-founded GMPVC, a media-for-equity fund, which allows media groups to use their advertising inventory to invest in startups. At IBM, he was responsible for the media industry in Europe, working on the digital transformation of key accounts. Niko is widely published. His latest book, "SimplySeven" (Palgrave Macmillan) is a guide to the seven business models of the web. Niko has a Ph.D. from the London School of Economics and also studied at the Johann-Goethe-Universität in Frankfurt and Brown University in Providence.

Tilman Rotberg is overlooking GfK's Media and Entertainment business in Germany. Before taking over this role, Tilman was the Global Industry Lead for Technology clients at GfK; among others he was responsible for building up GfK's Technology, Digital and Media presence at the US West Coast and opened up GfK's office in San Francisco/ USA, which he managed for five years. Tilman received his BS and Master's degree in Psychology from the University of Bonn, Germany.

Florian Renz is a Senior Manager at GfK Advanced Business Solutions department for GfK Panel Services Germany. Based on the single source approach "GfK Crossmedia Link" he is responsible for ad effectiveness studies for different media clients. Since 2008 Florian is GfK Key Account Manager for Facebook in Germany. He studied Sociology and Business Management with a core focus on Communication Sciences at the University of Bamberg, Germany and completed in 2015 his studies with a postgraduate degree in Interactive Marketing.

GfK is a globally trusted source of relevant market and consumer data. For the media and advertising industries, GfK provides data, enabling advertising sales in television, print, web and mobile. The views expressed in this piece are the author's own, not those of GfK.

How to Be a Successful Publisher in the Programmatic World

Frank Bachér and Jay Stevens

Within just a few years, automation has changed the face of advertising as we know it. The impact on publishers, advertisers and agencies is enormous, and on going.

Publishers have to deal with a lot of challenges: They have to find creative ways to monetise their increasing mobile inventory through new ad formats, to leverage data and new formats, not to mention hiring or retraining staff who can thrive in this shifting environment.

The option to sell premium inventory through automated guaranteed is another game changer. With automated guaranteed, reports suggest the promise is of another big chunk of digital advertising moving to being sold programmatically which results in significantly improved efficiency.

Defining programmatic targets for sales teams is a further logical step towards the future of automated advertising. Every publisher should also have an opinion on seller co-operatives (also known as alliances or coalitions): in many European markets, as well as worldwide, publishers have founded co-ops to increase scale and pool data around their unique, high quality audiences.

Buyers and sellers alike should be aware of the rapid development and evolution of advertising automation: third parties suggest that in future, there will be no 'programmatic', but rather data and automation will form an intrinsic part of all advertising.

F. Bachér (✉)
The Rubicon Project, Kurze Muehren 1, 20095 Hamburg, Germany
e-mail: fbacher@rubiconproject.com

J. Stevens
The Rubicon Project, 2nd Level, 25 Procter Street, London WC1V 6NY, UK
e-mail: jay.stevens@gmail.com

1 Focus on Mobile

All major publishers are currently facing the same challenge: they are all seeing a significant increase in mobile inventory. In fact, most of the big editorial content or e-commerce providers already have at least a 50 % mobile traffic share. However, monetising mobile inventory at the same rate and volume as desktop inventory has so far been a challenge. In fact, mobile advertising investment has traditionally been seen as pretty anaemic when compared to other channels – this is astonishing given the fact that already in 2014 internet usage on mobile devices exceeded PC usage.[1]

But the mobile marketing industry now seems to be reaching a tipping point: the traditional gap between mobile ad revenue generated and the exponentially increasing time consumers worldwide are spending on their phones is starting to narrow. This is due to a variety of factors, such as the increase in high quality mobile inventory, brands' renewed interest to better understand this channel, the growth in mobile bandwidth, as well as the increasing adoption of mobile native advertising.

1.1 Native Advertising: New Opportunities for Buyers and Sellers

A large opportunity for publishers to monetise their growing mobile inventory is to sell native advertising and high impact formats on their mobile websites or apps. The results of the latest Rubicon Project Mobile Buyer Survey[2] show that spend on mobile native advertising is set to more than double in 2015 globally as buyers around the world shift a greater share of mobile marketing budgets to these immersive formats. Agency trading desks predict that native advertising will secure a 13 % share of overall global mobile spend in 2015, up from just 5 % in 2014. That is a significant share because mobile native advertising wasn't even available to buyers before 2014.

The survey also suggested that the buying of automated guaranteed mobile advertising is also set to grow globally in 2015 as buyers increasingly look to automate future deals on a guaranteed basis as well as in real-time. Eighty-one percent of respondents noted that embracing automated guaranteed trading was an important trend for this year.

In any case, standard formats are still the most in-demand type of mobile advertising from buyers around the world (accounting for 46 % of spend), yet formats such as rich media (26 %), video (17 %) and native (10 %) are growing in popularity across the board.

---

[1] Comscore: http://searchenginewatch.com/sew/opinion/2353616/mobile-now-exceeds-pc-the-biggest-shift-since-the-internet-began.

[2] Rubicon Project 2015 Mobile Buyer Survey: http://www.prnewswire.com/news-releases/rubicon-project-releases-third-annual-mobile-buyer-survey-at-mobile-world-congress-300045921.html.

The survey also demonstrates that buyers are positioning themselves to not only move more of their budgets into mobile, but to also quickly gain expertise in mobile automation, with a focus on more innovative formats.

1.2   Standardisation Enables Scale

Standardisation is an important driver to better monetise your mobile traffic and brings scale to the business. Mobile monetisation via native ads is underdeveloped today and a key opportunity is to expand the reach into the programmatic world. The latest version of the OpenRTB protocol that has been released by the US IAB includes support for native advertising for the first time.[3]

This will give sellers and buyers the ability to buy and sell native ads in the programmatic marketplace, which could lead to a standardised way of trading native ads programmatically. Selling native ads programmatically will give publishers new opportunities to strengthen their mobile business and buyers will be able to reach premium audiences with high impact ads more easily. With the new potential for revenue, more app developers will work on new features and that will also attract buyers to invest in mobile marketing. Finally, by standardising the process and removing the need for the integration of multiple SDKs, automation promises to make ad supported mobile apps render more quickly and less likely to crash, thus delivering an improved user experience.

1.3   Mobile Bandwidth Brings Speed and Unfolds Creativity

The speed and capacity of mobile bandwidth is evolving, creating a perfect storm for a buoyant mobile advertising market, and 5G technology holds the potential to bring another leap on mobile usage. The possibilities this opens up for mobile video are manifold. Being able to access content at 5G speed and quality will enable mobile users to consume much more immersive content–drawing marketers in an increasingly compelling way because it finally provides them with a better canvas with which to tell their stories, in an intimate dialogue with the user.

This impressive growth in bandwidth does not just provide an opportunity for developed markets, but is also creating a truly global footprint beyond anything traditional media has ever been able to reach. Especially in African and Asian markets, mobile penetration is surprisingly high. In these developing countries mobile phone usage far exceeds desktop computer usage as a means to access the Internet. Mobile users are showing a strong preference for using their device for a variety of activities whose are normally performed on laptops, one example being mobile banking.[4]

---

[3] http://www.iab.net/guidelines/rtbproject.

[4] http://www.pewglobal.org/2014/02/13/emerging-nations-embrace-internet-mobile-technology/.

Mobile users in these countries might be relying on feature phones, but once they move onto smartphones in the not too distant future, it changes everything – mobile marketing will be able to access billions of people at once.

1.4 Wearables: An Incredible Opportunity to Engage with Consumers

After seeing the dramatic increase of mobile usage and mobile overtaking fixed Internet access in 2014, the question is "what is the next big trend in consumer engagement"? Some believe that wearable computers could be the next revolution while others are reluctant. Many are approaching wearables like the Apple Watch with much of the same scepticism that originally surrounded smartphones.

How could a small device like a watch be useful for anything? What content could possibly be consumed on a screen as big a thumb? And next to the consumer the major advertisers, publishers, application developers and analysts are thinking about business models for this new device. Advertising-fuelled revenues may prove attractive, as they have in the past, especially versus download price in the app economy.

1.4.1 A New Step in Location-Based Advertising

Mobile advertising has given marketers great opportunities to use location-based advertising: this is used to pinpoint consumer location anonymously and provide location-specific advertisements to mobile devices. With wearables, the promise is that marketers can even more precisely reach the user in a specific place than a mobile device or laptop. The laptop stays on the desk and the mobile device sometimes stays in the pocket, has to be charged or sometimes stays in the home. But the wearable goes absolutely everywhere, the gym, the morning run, the evening shop, and it stays on your wrist even at night. Such rich location data is powerful for advertisers.

1.4.2 Wearables Can Deliver Extensive Insights for Marketers

When Apple announced the Apple Watch, they called it "...The most personal product we've ever made, because it's the first one designed to be worn." The ability for this device to track your body habits is astounding, collecting around-the-clock bio data: pulse, temperature, calories consumed, steps taken. The watch is a device that can know and monitor human habit more closely than any other device has the ability to.

With wearables, health data in particular comes into play for advertisers. The device tracks how sedentary one has become in recent weeks and that could in theory be used to help encourage users to find the nearest running club or gym via advertising. Bio data being shared with advertisers is not something consumers have been widely exposed to so far, but the potential for this device to help live a fitter, healthier life is significant, and actionable bio data for the advertiser unlocks a tremendous opportunity.

Wearable developers must ensure that this extremely personal data is treated safely and anonymously, and only at the explicit consent of the user will anyone have access to it. It is also incumbent on the brand to build an environment in which users gladly share bio data to receive a better advertising experience. The requirements for brands are high: they must create a value exchange so compelling and so safe in order to get access to such personal data on such a personal device.

The challenge for publishers, advertisers and app developers is to develop new ways for brands to tell their stories to consumers. And the past has shown that publishers and advertisers will find creative ways to reach audiences on new devices. According to some experts, the biggest potential will be in superior location-based advertising with more personalised targeting opportunities in real time. Some of these opportunities may be delivered through automated advertising.

2 Data: A Publisher's Best Friend

Whether it's big, first party, or in the cloud, Data is without a doubt one of the biggest trends we currently hear about. Alongside automated advertising, data has to be one of the major talking points of the past few years for publishers and advertisers alike.

But, a few new buzzwords aside, is data really anything new in the media world? Publishers have had subscription (a.k.a. data) departments for decades, monetising their lists for both direct and email marketing.

However, as far as advertising goes, the debate certainly has advanced over the past year. Online advertising has seen more and more private marketplaces that include data, therefore unlocking a valuable asset that publishers have in the past been challenged to monetise.

But what is the latest thinking on data in advertising, both from buyers and sellers? Has the debate really moved on from the 'subscriptions' era, and where is innovation happening?

2.1 Seller Data Versus Buyer Data

First, it's worth outlining the differences between buyer and seller data.

For publishers, data means understanding the millions of customers' actions on a site or mobile application and what they mean from a product perspective. This data could be based around search, registration or even context. When it comes to advertising, the key is in understanding how to use that data to get to a product that's saleable and performs for advertisers, but is also relevant to users.

Specifically in terms of programmatic advertising, this means taking into account data from the ad server, automation platform and bid landscape report, from a high level to a granular one – advertiser-by-advertiser.

From the buyer perspective, actionable data could be as simple as what is contained in a bid request, what data is attached to that bid request, or what inventory it is appended to.

Essentially, it is about discerning which inventory works best, with what data coupling, and which impressions are performing best at what time of the day.

The buyer's view of data is clearly very different to the seller's. However, where the two datasets intersect, whether on desktop or mobile, is where publishers are able to monetise it - passing first party data safely within private marketplaces.

2.2 Programmatic Advertising Driving Relevance

Some publishers have been using audience data in advertising for quite a while, but automated advertising makes that data even more relevant – quite simply because the audience structures are created in real-time. In fact, the data is so recent that someone can be included or excluded from a segment in one user session.

If a manufacturer wants to target people who are interested in a specific product, it is incredibly powerful to be able to reach that group in real-time. Still, it's worth clarifying that this type of targeting is just a part of the chain – though an incredibly effective one for both buyers and sellers. One highly targeted private marketplace would normally form part of a wider campaign – you still need to find numerous touch points across different screens and devices – which in itself is something automated advertising is increasingly enabling.

2.3 Reaching Scale Through Data

This is where having a rich understanding of the bid landscape comes in; the publisher can help feed information through to buyers on what bids they're winning, and get their feedback on what works – all of these steps can help structure the deals that publishers make available on private marketplaces, going from a very broad to a very granular level at different price points.

On the back of understanding how these perform, publishers can then look towards having a pricing model which places more emphasis on supply and demand.

At the moment the industry is still just making its first forays into first party data in automated advertising - to reach greater scale, publishers and buyers need to think about how they can extend those audiences and amplify them across a range of marketplaces and even beyond them. From the seller's perspective, tactics like audience extension where the publisher uses technology to sell its audience both on its own sites and mobile applications, as well as other publishers' sites, is an important factor.

2.4 Future Growth Drivers

For data-driven advertising to continue to evolve and grow this market, there is further work, which needs to be completed, including:

1. Finding even greater scale for buyers, while maintaining quality
2. Increasing automation around the edges
3. Improving how private marketplaces are packaged

One final consideration, which is key to the future, is also driving brand spend to this channel.

It has been suggested that algorithms may still have to evolve somewhat before programmatic advertising can fully measure and optimise to brand metrics. However, if anything, the growing debate around ad viewability, and the adoption of new 'in view' metrics will assist.

The next step for programmatic advertising will be in determining how those brand metrics are standardised and applied universally to the automation space, as well as the wider digital market.

3 The Next Step in Programmatic: Automated Guaranteed

The advertising industry has seen the rise of real-time bidding and private marketplaces over the past years. However, these two segments are potentially overshadowed by the market for guaranteed orders, where inventory is still largely bought and sold manually. This significant portion of the digital ad spending couldn't be addressed through ad technology in the past. The latest evolution of automated advertising is to deliver these guaranteed deals programmatically. With this development, the promise is of a significant part of the digital advertising market moving to programmatic. With automated guaranteed (also referred to as programmatic guaranteed or programmatic direct) the efficiencies of automation can finally be applied to the entire range of deals and campaigns buyers and sellers can execute.

Ultimately, automated guaranteed allows publishers to sell a specific number of ad impressions for a predetermined price over an agreed amount of time and buyers to access inventory on the same upfront basis.

3.1 Reaching a New Level in Automation

With automated guaranteed the digital advertising business is reaching a new level: Every publisher can sell his entire inventory programmatically, non-guaranteed through real-time bidding and private marketplaces, as well as on a guaranteed basis through automated guaranteed. The ability to apply automation across all inventory helps publishers to streamline the sales process further than ever before, saving time and costs both for themselves as well as buyers.

A recent report from IDC revealed that the global market for guaranteed orders could reach $41bn by 2019.

The rapid growth of private marketplaces around the world is a key indicator that sellers and buyers alike have a strong appetite for further efficiency. Advertisers meanwhile realise it can mean that more of their budget goes towards the media rather than it going to operational overheads.

3.2 Programmatic Targets for Sales

Many publishers started programmatic advertising a few years ago with very small teams. Sometimes a junior member of staff was the first who was responsible for setting the parameters on a sell side platform and to pull reports for the sales director. In the last 2–3 years, publishers have since created yield or performance optimisation teams, recruiting experienced data analysts or yield managers after seeing the increasing revenue coming through automation. However, in the first instance there were few defined sales targets around it and the traditional sales team was not involved in the growing revenue that came from programmatic advertising.

With the increase of private marketplace deals, and especially the rollout of automated guaranteed, every sales person can now sell campaigns programmatically. With this latest development, every sales person now needs a programmatic goal, as it is insufficient to have sales goals for the operational programmatic team only.

4 The Rise of Programmatic Co-operatives

In a number of European countries the rise of publisher coalitions or co-operations is an increasingly common trend. Publishers in countries like Czech Republic, Denmark, France, Italy and Greece have already agreed to combine their collective inventory and data. There are also signals from other markets, both in Europe and worldwide, to scale their business through embracing programmatic alliances.

The latest attempt to satisfy advertisers' desires to reach quality online audiences is the programmatic alliance Pangaea. Five news publishers, The Guardian, CNN International, the Financial Times, Reuters and The Economist formed Pangaea to allow marketers to use programmatic buying technology to purchase advertising space across their inventory through a dedicated marketplace. Another reason for bundling the online properties is data. The combined audience data gives marketers the opportunity to target specific types of consumers across the properties of the member companies. The five publishers can share first party data with each other and sell unique audience segments with a combined reach of 110 million unique users.

4.1 Europe Is Leading the Development of Co-operatives

As far back as 2012 and 2013, publishers in Europe started to embrace alliances like Dansk Publisher Network (Denmark), CPEx (Czech Republic) and La Place Media (France). Publishers were motivated to bundle their online properties programmatically due to the size of the markets. In many European countries the number of Internet users is around ten million or lower. In these markets it is more expensive to run a sales and operational team, with a lower revenue per capita and lower margins.

Global players like Google and Facebook don't have this challenge. They collect data from hundreds of millions of Internet users and can scale their business without being constrained in the same way. It is difficult for traditional publishers to be able to compete against such global competitors. Leading publishers might have meaningful size and reach in their respective markets individually, but they still may not be able to sufficiently scale their advertising business through data. A competitive advantage of the co-op is the quality of the portfolio. In all of the European markets where co-ops were founded, only premium publishers with strong brands and a leading market position participate in the alliances. The combination of the quality of the brand and audience, plus the higher combined reach is what buyers can expect from co-ops.

One example of a successful programmatic co-operative is the joint venture La Place Media. La Place Media has been operational in France since 2012 and it includes 250 publishers, including Amaury Médias, FigaroMedias, Lagardère Publicité and TF1 Publicité. Five of the biggest publishing houses bundled their online properties, offering a premium portfolio and audience through automated advertising. The motivation for this alliance was again size and scale. Supply was too fragmented and even the biggest publishers in France only had the reach of a medium-sized publisher in the US.

The results are encouraging: La Place Media reaches 70 % of the French Internet users and delivers four billion impressions per month. CPMs of La Place Media jumped 70 % in its second year of operation and the trend is continuing. "2015 will be by far our best year, from starting small, we will do 20 million Euros revenue" says Managing Director Fabien Magalon.[5] "La Place Media didn't develop their own technology, they co-operate with the leading ad technology companies". Fabien Magalon explained that "these media companies don't truly have technology in their DNA, so we license multiple third parties in order to be able to operate our strategy."

The five shareholders of La Place Media already have a common understanding of the next steps: "There is a big expectation for publishers for a unified platform that combines programmatic, guaranteed and programmatic non-guaranteed, and web, mobile and video," according to Fabien Magalon. There's also the opportunity

---

[5] http://adexchanger.com/publishers/how-french-publishers-reclaimed-programmatic-by-creating-la-place-media/.

to duplicate the idea of La Place Media in the offline world, building a marketplace for programmatic print.

La Place Media is a great example of how a publisher co-operative can increase revenue, how the idea of an alliance can be extended to build a unified platform and to include other media. The next question is: will there be larger Europe-wide alliances in the future after programmatic co-ops have been successful on a country-by-country level?

Frank Bachér As Managing Director of Northern Europe for Rubicon Project, Frank Bachér is responsible for selling the automated advertising platform to buyers and sellers in the DACH region (Germany, Austria, Switzerland), the Nordics and Eastern Europe.

Before Frank Bachér joined Rubicon Project, he was Vice President Online and Central Europe at travel booking technology company Sabre Travel Network. He spent 3 years at Germany's biggest digital sales house InteractiveMedia CCSP GmbH, where he was responsible for marketing and sales. In this position, he also drove innovations in the fields of mobile, IPTV and yield optimization. As Chairman of the Board at eBay Advertising Group GmbH, he was responsible for all online sales activities and partnerships of eBay in Germany and the classifieds portal mobile.de. Additional tenures include Tomorrow Focus AG, Verlagsgruppe Milchstraße and Axel Springer Verlag.

Frank Bachér is also a longtime member of Bundesverband Digitale Wirtschaft (BVDW) and Verband Deutscher Zeitschriftenverleger (VDZ).

Jay Stevens Charged with spearheading the company's international expansion efforts, Jay brings more than 15 years of interactive marketing and international business experience to his role. He has been responsible for building the company's presence from the ground up across the UK, France, Italy, Germany, LATAM, Japan, APAC and Australia.

Before joining Rubicon Project in May of 2009, Jay served as SVP at MySpace/Fox Interactive Media. He was the social network's first hire outside of the US, where he launched and oversaw operations and expansion across 12 European territories.

Prior to joining MySpace in February 2006, Jay served as Director of International for Silverpop, the leading email marketing software and services provider and greenfielded their business into the UK. He previously was a member of the founding team and served as the Director of Marketing for Radical Communications, where he drove the company's day to day marketing operations. Jay began his career in interactive media in

1998 directing the digital marketing practice for AlexanderOgilvy Public Relations, managing the DoubleClick and RelevantKnowlege accounts for the boutique technology marketing communications firm.

Jay has been recognised by the Evening Standard as one of the 1000 most influential Londoners, by Revolution Magazine as the 5th most influential person in digital media in the UK and by Advertising Age as one of the top 20 interactive marketers of the year in the United States. He has also served as a press aide to former President Jimmy Carter and holds a B.A. in History from Emory University.

Part III

Transformation

The CMOs Challenge

Ralf E. Strauss and Jonathan Becher

Programmatic Advertising offers the ability to provide individuals with authentic, personalized content – and to do so in an efficient way that minimizes budget. As a significant evolution in online marketing, Programmatic Advertising relies on creating more robust user profiles based on first and third-party data and pairing cookie data with log-in and CRM information. Like other digital transformations, Programmatic Advertising presents a variety of challenges to the CMO, such as building up the necessary skill-sets, change management to support the transformation, and the integration into existing processes and IT landscapes. Outsourcing provides some short-term advantages, but as programmatic spend increases, internal structures and competencies will be required.

1 The State and Nature of Programmatic Advertising from a CMO Perspective

Programmatic buying is heralded as the backbone of the future of advertising. Programmatic covers a wide range of technologies that automate the buying, placement and optimization of advertising, thus replacing human-based methods and processes. Through programmatic technologies, advertisers can buy ads the same way they pick up something on *Amazon* or bid on *eBay*. Estimations as of today go as far as to project that programmatic spending will reach ca. 20–25 % of the overall digital-ad market in the majority of Western countries already in 2015.

In part, programmatic is attractive because it can save money: through automation, transactions become more efficient, cutting out complex ad-operation tasks

R.E. Strauss (✉)
German Marketing Association, Huusbarg 40, 22359 Hamburg, Germany
e-mail: ralf.strauss@customerexcellence.de

J. Becher
SAP, 3410 Hillview Ave, 94304 Palo Alto, CA, USA

and provide transparency about pricing. Conversely, it also requires further spending for tech, optimization, data, higher skilled staff etc. As IAB stated, media today is almost infinite, money no longer guarantees success and it's not enough to just be seen (IAB 2014). Brands have to be relevant and presented in context to have any chance of cutting through the clutter and engaging their audience. Digital advertising and targeting allows brands to reach and engage individuals like never before. Major changes in the media landscape have supported and enhanced this ability to understand and respond to buying signals and drive advertising effectiveness, such as (IAB 2014):

- Advertising is no longer dependent on media as a proxy for audience – but can be bought and sold at an impression level basis, targeting single users;
- Social media is no longer free, with new advertising formats continually emerging to take advantage of its inherent ability to sustain reach;
- Mobile advertising is growing, bringing new opportunities to market;
- Measurement and analytics are improving across different channels;
- Organisations are accelerating to real-time consumer engagement – pushing further to permanently reconcile and evolve marketing plans, budgeting, activities and – consequently – organizational structures within a broader context of digital transformation.

In this vein, the programmatic share of display and video is increasing rapidly, also re-opening traditional markets like CRM, which will be directly linked into programmatic platforms to establish one integrated process from marketing planning and program definition, to the setting of KPIs at the campaign management level across different communications channels (Strauss 2008). Since 2014, *Facebook* already offers Marketing Partners (Technology Companies) to use their API access to build custom audiences based on CRM data. Closed feedback loops based upon real-time dashboards allow the intermediate reconciliation and optimization of all tactics and activities. The flooding of new DSP and SSP technologies paired with unprecedented volumes and quality of data enables authentic content to be presented to individuals, in real-time. The new programmatic platforms therefore provide more robust and centralized user profiles based on first and third-party data and pairing cookie data with log-in and CRM information for the right context. They are able to do so with minimum data and information loss, and can apply semantic analytics (Fig. 1). As a consequence, advertisers and publishers quickly learn the advantage of programmatic and by now programmatic adherents outnumber those in favor of discretionary inventory.

Additional growth to the programmatic segment will be fueled by the continuous advance of online, mobile, video and digital media, the shift of classic CRM-budgets into programmatic (to transform into "User Relationship Marketing"), and the switch of branding budgets to programmatic premium. Many media agencies have pledged to automate 50 % of their clients' media buying by 2016. It is clear that in the future the majority of buying will be done programmatically, yet

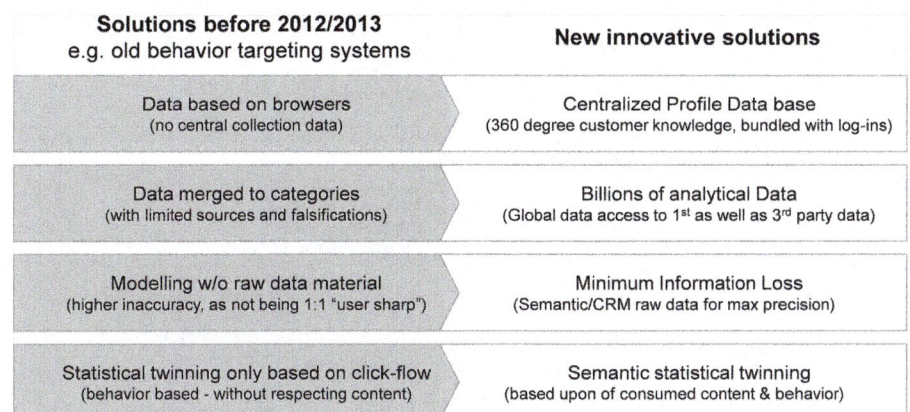

Fig. 1 Developments in programmatic technologies (Source: United Digital Media 2014)

there are always exceptions to the rule. For example, deep brand integration, content plays and tent pole events cannot yet be accomplished programmatically.

In sum, three driving forces are behind the rising success of programmatic advertising.

- Ad delivery robots maximize process efficiency, when online advertising inventory is bought and delivered user sharp at predefined terms and conditions. Direct technical interfaces ("private exchanges") needed between publishers and advertisers substitute existing media value chains;
- With user sharp ad inventory, data is key for an efficient user selection. Besides the dominant online pure plays such as Google or Facebook with massive amounts of data, third party data pools emerge – or private cookie pools are built up by advertisers. Not only for new customers, but also for implementing a cookie-based "User Relationship Management".
- Innovative hashing tools yield superior data quality by combining hard facts (log-in data) with the explosion of first (owned) and third party online data. The result will be such a critical lift in advertising performance that pure online data players are forced to strengthen their service offerings in response. While over the last ca. 3 years, ca. 2 bn € have been invested in ad tech globally, we now can rely on proven platforms and systems. User sharp ad delivery plus more and better data and inventory (non-remnant) is yielding superior performance, while the upcoming connection to "hard" data quality (log-in) from traditional CRM systems will yield even more performance. The same is true for video: here in particular *YouTube* will be a driving force in the market.

Consequently, programmatic is no longer used for remnants only, but is an effective way to increase E-CPM in discretionary inventory as well. Now that it is clear premium prices can be applied based on the added-value of the data and content, publishers will continue to open inventories for programmatic purchases

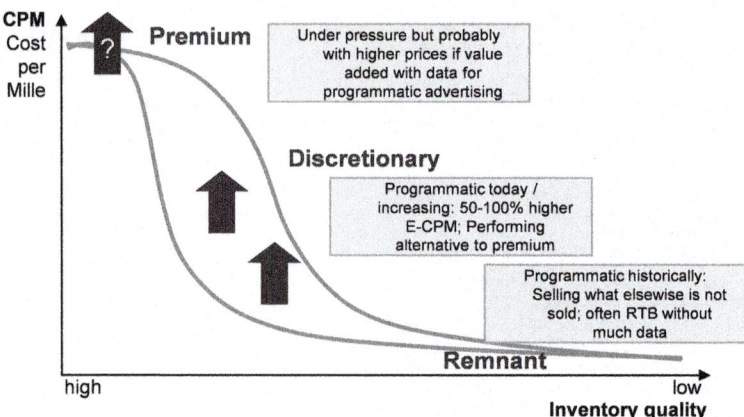

Fig. 2 Inventory quality versus CPM in programmatic advertisements

(Fig. 2). Therefore, the future in programmatic will be about eco-systems based upon data and media, including the so-called "premium" as well.

Even premium will be sold programmatically soon: forced by the tremendous data quality and closed data-media-solutions (eco-systems) by *Google* and others, as well as by big advertisers, it can be expected that publishers will open premium stepwise to programmatic. This will happen in private exchanges, which are going to be established without transparency towards third parties, like agencies – and consequently probably lead to a bigger market share of programmatic than expected as of today. In addition, publishers are already trying to build up proprietary data-media-solutions (eco-systems) by themselves in order to capture the benefit in terms of margins of data-driven media, instead of pure (lower value added) advertisement bookings.

As of today, while international giants like *Google, Facebook* or *Amazon* dominate the market, their objective is to catch as many users/advertisers as possible – sometimes with offering additional services for free – that can either deliver ads or collect data to improve targeting. Valuable customer insights are directly utilized to improve respective media offers and random inventory is sold via the *Google* network. From their perspective, their growth strategy is focused upon catching big branding budgets, provide market leading technologies (DSPs, RTB), offer random & remnant premium inventory, and offer cross device products (display, video, mobile, etc.).

In this vein, the CMO of *Procter & Gamble, Marc Pritchard*, already predicted the end of mass media: *". . . continuous change is our new reality. We will never again have the opportunity to reach all customers via one single platform . . . we continuously improve efficiency of our marketing spending by strengthening our digital activities as main driver of ROI . . .".* As one consequence, *P&G* is said to plan to operate 70–75 % of its digital spending via programmatic media, so that ¾ of their display ads will be programmatic (while *American Express* and other big players already have stated similar objectives). One of the bigger hurdles still is to get enough programmatic inventories for higher valued brands, as low quality remnant inventory could jeopardize the high quality brands.

Saying this, also the traditional relationship with media agencies for the ad buying process comes under scrutiny. While media agencies initially combined functions like the development of a media plan, the consultation of the customer or the actual media-buying function (at high direct or retainer based margins), now machine-based purchasing takes over. How will the role of the media agency evolve in future? All interviews indicate that there still is a need for a media agency, but with different functions and roles – migrating from the predominant media-buying function into a more consultative role, while the purchasing, operations and execution will be machine based.

2 The Challenges From a CMOs Perspective

A vast amount of interviews, roundtables, discussion sessions with more than 60 CMOs within the CMO Community since 2014, as well as some recent research indicates, that CMOs are being confronted with at least nine different challenges to move into Programmatic Advertising (Fig. 3; Peterson and Kantrowitz 2014; Strauss 2013; IAB 2014).

First of all, the shift into programmatic involves a massive **change** as well as **knowledge transfer** and change in required **skill-sets**. E.g. in 2014, the Association of National Advertisers surveyed 153 marketers in the US and found that only ca. ¼

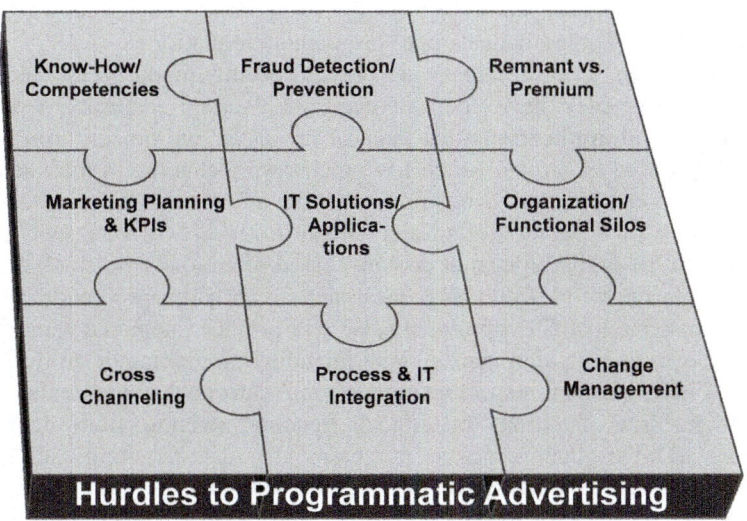

Fig. 3 Hurdles towards programmatic advertising (Peterson and Kantrowitz 2014; Strauss 2013; IAB 2014)

of respondents said they knew what programmatic buying is and have actually used it (IAB 2014). The other three quarters ranged from either being completely oblivious, aware – but unclear or in need of education. But even those who use programmatic tools to buy media may not have a firm grasp of what that means in all extensions.

Secondly, digital advertising is said to provoke a **fraud** problem. Much of the fraud is occurring in the programmatic ecosystem, where an opaque marketplace is (at least) able to allow insidious actors to list networks of shell websites, flood them with non-human traffic and cash in order to defraud advertisers. Even whilst the digital-ad industry reiterates its commitment to solving the problem, nearly every entity along the supply chain in Programmatic Advertising benefits in some way from the fraud, or – to be politically correct – at least has little incentive to stamp it out by definition (Peterson and Kantrowitz 2014). Publishers make money from it, buyers' performance looks great, and technology companies get paid to help stop it. In most cases, brands end up holding the bag and pay the bill.

Thirdly, it can be expected that **premium publishers** will also go programmatic over time. So far, publishers have very good reasons to be wary and stay out of the programmatic game. If media buyers embrace programmatic buying simply as a cheaper way to buy ads, premium media sellers have understandably little incentive to join the bandwagon (Peterson and Kantrowitz, 2014). Premium publishers such as *The New York Times* or *The Wall Street Journal* have set up private ad exchanges as a way to dip their toes into the automated selling of their inventory. As one consequence, publishers have begun to automate their direct dealings with advertisers. Under these "programmatic direct" deals, a publisher's sales rep may negotiate an arrangement with an advertiser that includes top-tier inventory like home-page-takeover ads at a fixed price for a guaranteed number of impressions. In this vein, private deals appear to be the biggest opportunity and direct way to easily migrate the vast of brand budgets onto programmatic platforms.

Fourth, another set of dominant players will occupy market positions, such as *Google*. *Google* does have the dominant ad "stack" – means a **series of technologies and applications** that manage the ad-buying process. But there are others like *Yahoo!* or *Adobe*, which have their own technologies that manage the process end-to-end like "all-in-one programmatic ad factories". It can be expected that *Facebook's* Ad-Server *ATLAS* and its SSP *LiveRail* will very soon close the loop, to form the first profile- (not cookie-) based programatic ad-stack that works in a cross-device world. Following discussions with software vendors, it can be expected that classical CRM providers also will provide integrated solutions from planning down to campaign management including programmatic ad delivery. As of today, CMOs and their respective organisations can cobble together their own set of solutions from hundreds of ad-tech vendors, ranging from demand-side platforms (DSPs) that process purchases to data-targeting solutions to ad-verification services that make sure the ads were actually seen by real people (Peterson and Kantrowitz 2014).

Fifth, while the technology stack will further evolve, more and more CMOs are looking for a direct **integration** into their existing processes and IT applications.

Saying this, it can be expected that after the first period of having green-field applications or outsourcing all operations e.g. either to media agencies or specialised trading desks, the need arises for a direct integration e.g. into marketing planning, campaign management, or the financial integration for the settlement of all payments with classical ERP systems.

Sixth, all digital advertising work's best when it is aligned and integrated to the wider **marketing planning and marketing mix**. E.g. from an advertisers perspective, being accustomed to focus on classical GRP measures, now being confronted with a different set of KPIs – e.g. indicating that a million people saw the ads, but with no indication on which sites and in which context environment. A situation which is pretty contradictory to the old habit of using media and delivering ads in clearly defined environments and contexts. So, while programmatic is often siloed in the "test & learn" phase, it should be reintegrated as soon as possible as part of the wider marketing strategy and according plans. Interviews indicate that there is tremendous potential for programmatic to help brands adapt more quickly and flexibly in a real-time, multi-screen world. More and more, a proper marketing planning focused on different target segments and cascaded from strategy down to tactical levels is required, based on consumer and market insights and ammunitioned with target-group specific offers (Strauss 2008). As programmatic continues to grow and becomes the enabling technology for all kinds of media transactions, it will cease to exist as a line item where specific budget allocation is required and is said to become instead the de-facto method by which the whole plan is executed (IAB 2014). As a result, budget will no longer be allocated to programmatic initiatives but rather the whole plan will be delivered programmatically. In this vein, the budgeting process and budget allocations need to be reconsidered, moving from fixed budgets, based on historic data into a "smart budgeting" process.

Seventh, the more programmatic methods take over, paving the way into a true real-time marketing, the more the classically functionally organized marketing organization with (specialized) silos in CRM, Online, Media or Customer Analytics comes to an end. Saying this, a closer collaboration in cross-functional teams is required – e.g. using agile methods such as scrum – bundling heterogeneous competencies and experiences across organizational silos.

Last, but not the least, as more TV and movies are delivered over the web, programmatic technologies will take hold **across channels**, too (Peterson and Kantrowitz 2014). E.g. *Hulu* has experimented with auctioning off video ads through a private exchange. And *ABC* announced that it would let some advertisers buy ads against *ABC's* digital content programmatically. Even some traditional TV ads are getting automated. Cable carriers such as *Comcast* enable automated ad-targeting against on-demand videos. And satellite carriers *DirecTV* and *Dish Network* sell some TV inventory programmatically. In sum, digital delivery and data is driving the usage of programmatic. It can be expected as a next step in future, that out-of-home, digital television and radio will follow, as being an integral part of future mobile marketing scenarios.

3 Managing the Change Process into Programmatic

With programmatic being a disruptive force in traditional media markets, it seems obvious that the change process is accompanied by uncertainties. Similar as in outsourcing, CMOs report back the need for change – the underlying root-cause across all organizations: one fears allowing something or someone else to do their job will cause theirs to be dissolved. There are usually complex reasons for a resistance to change. At the employee level there are phenomena like (Strauss 2008):

- old (comfortable) habits such as working together with media agencies;
- selective processing for information that does not fit in their existing frame of reference, such as using KPIs outside the traditional and accustomed GRP framework;
- a high degree of dependence on the values, attitudes and beliefs of their most important contacts internally, as well as with media agencies and publishers;
- insecurity and regression, such as the fear of losing their job or the danger of de qualification due to a lack of understanding of programmatic – even whilst having been an expert in media for many years;
- a socio–psychological fear of the "new" and the "unknown".

Accordingly, a change process of this type like in other projects in Digital Transformation is subject to considerable fluctuations and stages in most organizations (Fig. 4). After a stage of shock upon receiving information about changes that will take place in media, there is a stage of denial, which is followed by a stage of gradual realization and insight, until the planning schemas and implementation processes are ultimately accepted. The changes should be accompanied

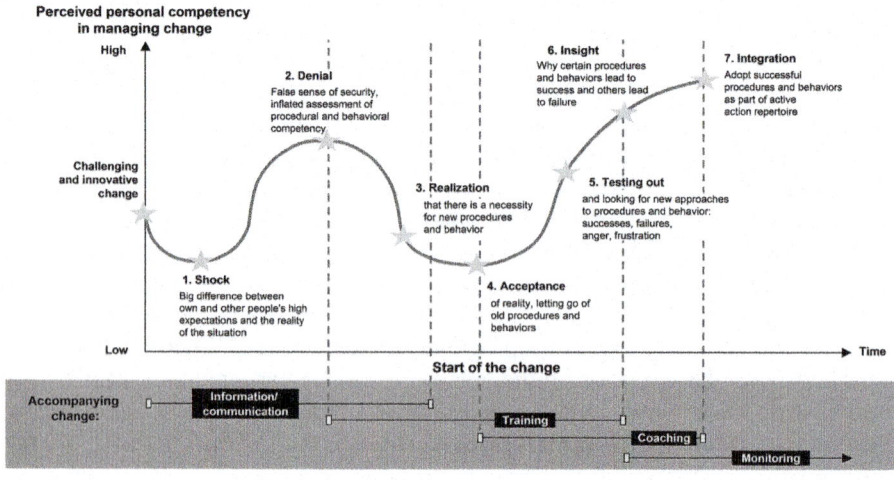

1. 4 Change management cycle for implementing change

by comprehensive change management that employs different tools that fit with each stage.

It is normal to fear the unknown, but in order to remain competitive, it is important to remember that change is the norm. Instead, reality and projects demonstrate that by welcoming programmatic, buyers can focus more on strategic decisions, target strategies and offers associated with the ad buying execution, rather than the routine aspects of buying that can be calculated by algorithms and the manual handling of media assets. Ad buyers can best secure their future in the business by accepting change, learning as much as they can and developing their skill set to include programmatic competencies. Programmatic will support this basic purpose of doing business by creating a more fine-tuned way to complete the task of ad buying on the basis of user-sharp profiles, while freeing up the buyers valuable time and resources to focus on the less predictable, more human areas of the job, such as translating consumer insights into targeted offers which will be appreciated by the individual target segments.

4   How Critical Are Technical and IT Skills in Marketing?

If we go back into the history of marketing, it can be described as a story of gut-feel and liquor-fueled inspiration. At least, that's how all the old assumptions of marketing might've rolled up into one general scene. Today, though, marketing looks decidedly different. Where creative craft was the old hallmark of the marketing trade, for the new school of marketers, technical agility and acumen based on data seem to rule the day. Therefore, the number of technical professionals working in the service of marketing is clearly on the rise, sometimes called "Data Scientists", "Programmatic Analysts", or "Marketing Technologists".

From a CMOs perspective, just having a technical staff is not enough. Marketing managers need to be ingrained to the level of understanding capabilities on different layers, in order to integrate e.g. Demand Side Platforms (DSPs) into existing processes and application landscapes. As in other marketing processes and applications, the discussion needs to be differentiated at least into four different levels of a business and IT integration scheme (Fig. 5):

- The (1st) Business Level examines how value added (products, services) arises between various business partners. Modern technologies and platforms helps to generate products and services more efficiently across several divisions and companies in the case of dispersed value added;
- The (2nd) Process Level illustrates the operational processes: at least one process is integrated between two or more divisions, departments or the like with one or more business partners, such as agencies or trading desks;
- The (3rd) Application Level outlines how these processes are supported by the information system: information system functions or data are accessed or used by another information system or user;

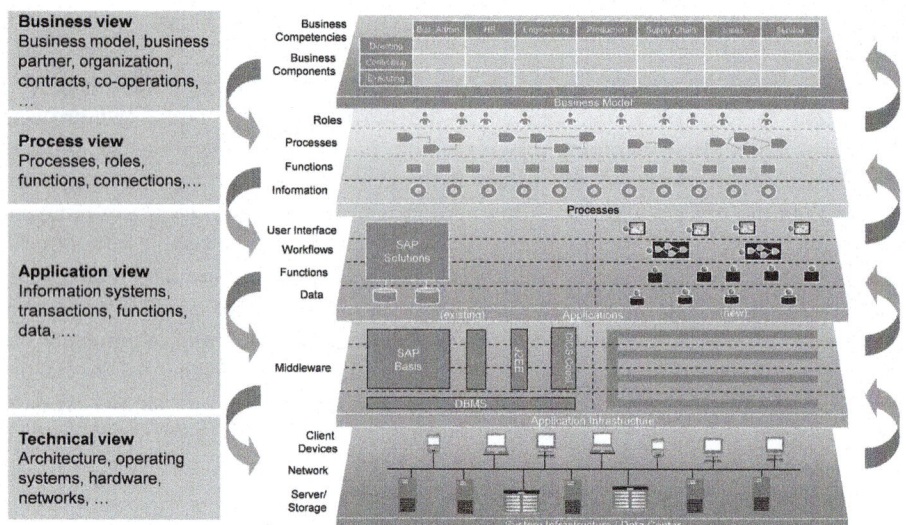

Fig. 5 Integration levels in digital transformation (Strauss 2013)

- The (4th) Technical Level considers the fundamental technical architecture and its components: integration takes place between two or more systems (internally/externally). Connection can be either synchronous (in real time) or asynchronous (batch).

As interviews and discussions with CMOs indicate, competences in marketing are needed on the business as well as the process layer. Furthermore functional know-how is valuable, e.g. when it comes to the application layer, while the pure technical view starting from middleware needs not to be covered and has to be located within the IT department. In essence, besides a functional understanding of IT at least on the application layer, programmatic it is not a question of technology, rather than a question of knowing the channels, mediums and moments that are best for reaching a specific audience on level 1 and 2.

A savvy marketer will be up to date on the value of technology and analytics for social monitoring, advertising and messaging optimization, but will also be aware of the right technologies for the right messages. What technology in general and programmatic in particular does is to force marketers to have more pointed and targeted messages to specialized audiences in order to increase conversion, not just reach. The mantra is not just to be able to offer more than just the value proposition to the consumer, but to offer how that particular value proposition specifically fits the needs and goals of the individual consumer in a specific context. Technology and analytics refine that process as an enabler. Experimentation in programmatic comes from having the right tools, asking questions, testing a hypothesis, adjusting levels in different marketing channels, and reviewing the impact on the verge of level 2 and 3. With all these levers, channels and degree of automation, having tools is the only way to consolidate and get a grasp on this information. As projects indicate, major problems pretty often occur already on the business process layer, where only unprecise descriptions and know-how is available. This sort of half-minded business process know-how then easily gets mixed up and confused on the lower technical levels.

To turn the layer model the other way around, it is important to start with the business vision and processes as opposed to getting distracted by the so-called "bells and whistles" of technology. The goals of the business have to drive the strategy ... not the latest and greatest widget. Otherwise, as it has been put in some projects: you can easily end up with a *Porsche* when you really need a pair of sports shoes. At the bottom-line, technology and marketing used to run in different lanes, but today it's a relay, and the best CMOs have at least a close and collaborative relationship with their CIO counterparts.

The question still remains valid, where these previously nonexistent marketing technologists or business architects at the interface between marketing and IT should come from? Many might migrate from IT, where a subset of those professionals are eager to apply their technical talents in the pursuit of more exciting, customer-facing innovations that are recognized as driving revenue, not merely containing expense.

5   Outsourcing or Insourcing Programmatic Advertising?

As programmatic media buying gains traction, brand marketers – even those who have already adopted some model of programmatic – are working to determine which organizational model of programmatic will best meet their specific needs. They are choosing between different models, reaching from agency-managed, programmatic-partner-managed, down to self-managed. Some of the big brands benefit by taking their programmatic media-buying in-house. The reason: they try to manage the entire process inhouse and start building up competencies right from the start. While those benefits are real and significant, they also carry over to managed programmatic services, similar as managed applications in IT (Musumano 2014).

With regard to data security, programmatic partners in general claim to provide safe repositories for client data, whether the client relationship is through an agency or brand-direct. In most cases, self-managed solutions are developed and hosted by the same vendors that provide partner-managed solutions so, by definition, the security levels will be the same. In fact, rather than pulling first-party brand data into their own systems, partners can work directly from a brand's DMP, targeting certain consumer segments without actually transferring and onboarding the data.

On the side of data handling efforts, programmatic partners are said to support brands and advertisers to incorporate first party data into campaign targeting, but they also can seamlessly integrate other sources of data such as third party and proprietary into their targeting (Musumano 2014).

Additionally, an outsourced partner can be given access to a brand's site visitor data in order to optimize campaign targeting just as effectively as an in-house team.

By doing this, the outsourcing partner serves ads to consumers that match the persona of the highest-performing web visitors, thereby jumpstarting targeting at the beginning of a campaign, and adjusting audience targeting in real-time throughout its flight.

With regard to the effectiveness in programmatic, outsourced vendors certainly have a greater focus on on programmatic, 24/7. Brands considering in-house as their future model normally consider beginning with an experienced partner that can navigate the risks and pitfalls in the complex world of programmatic to get kick-started. On a longer perspective, transparency and insights into financial mechanisms and customer demands will require an insourcing again. As experience shows, an expert partner will certainly be more effective than the brand at least in the beginning. Depending on the form of cooperation as well the contract, an outside expert working as an implant can help brand marketers develop their own expertise short-term, while not sacrificing safety and efficiency at the beginning – and then later over time insource again to gather and expand the necessary know-how.

6 Outlook: The Future of Programmatic Advertising

Ironically the next big thing is likely to be something as old as the media industry itself. Trusted relationships and premium inventory were the initial foundation of the advertising industry, but have not been part of the initial move to programmatic. Now as the technology matures and market penetration increases, these traditional relationships will be programmatically enabled to the benefit of both buyer and seller. The arrival of programmatic direct deals means that the efficiency and effectiveness of programmatic technology can now be applied across the full spectrum of digital advertising. As an example of the potential for programmatic beyond real-time bidding, deals between some premium publishers and advertisers are already closed within one phone-call. Without programmatic, the buyer and seller may never have found each other and when they did, such a deal would have taken days or weeks to go live (Bentley 2014).

From the publisher's perspective, it might be beneficial to pool programmatic platforms in order to generate greater scale for premium private marketplaces. There are already some good examples of what publishers can achieve when working together to deliver premium content at scale. *Audience Square* has done this in France and last year some of Australia's largest premium content producers announced that they are working together on the APEX premium mobile exchange, to give advertisers a brand-safe, programatic mobile offering at scale. The right model for today's publishers is driven by their own programmatic strategy and the dynamics of their individual markets.

From the CMOs perspective, concerns around data security and privacy will remain one of the bigger obstacles, and solutions will depend at least in part on governments and legislation to address the issues globally. More immediately for the CMO, in an auction and data driven media world, advertisers need to bring their proprietary data from on-site and CRM into the auction game. Every advertiser who bids on exactly the same profiles as his competitors, is doomed to fail by definition. To compete, advertisers need to invest into their own technologies and internal know-how over time. Adservers, DMPs and DSPs will be part of the regular marketing technology stack along with content management systems or e-Commerce platforms. Digital transformation will require an overall understanding of brands, consumer insights, advertising impact, and audience planning. Complete transparency across customer journeys, user data and ad delivery, augmented by tools for planning and steering, will be at the heart of the marketing core process.

Bibliography

Bentley, M. (2014). The future for programmatic advertising in 2015. In 12Ahead, Dec 4, 2014, http://www.12ahead.com/future-programmatic-advertising-2015
IAB Europe. (2014) *Why and how 'programmatic' is emerging as key to real-time marketing success*. London.
Lieb, R. (2011). *Content marketing: Think like a publisher – How to use content to market online and in social media*. Indianapolis: Que.
Minelli, M., Chambers, M., & Dhiraj, A. (2013). *Big data, big analytics: Emerging business intelligence and analytic trends for today's businesses*. Hoboken: Wiley.
Musumano, E. (2014). 5 Benefits of outsourcing programmatic. In Choicestream, Oct 23, 2014, http://www.choicestream.com/2014/10/23/outsourced-programmatic/
Peterson, T., & Kantrowitz, A. (2014). The CMO's guide to programmatic buying. Nine things every advertiser should know. In Adage, May 19, 2014, http://adage.com/article/digital/cmo-s-guide-programmatic-buying/293257/
Strauss, R. E. (2008). *Marketing planning by design: Systematic planning for successful marketing strategy*. London: Wiley.
Strauss, R. E. (2013). *Digital Business Excellence. Strategien und Erfolgsfaktoren im E-Business*. Stuttgart: Schaeffer-Poeschl Verlag.
United Digital Media. (2014). *Developments in programmatic technologies*. Hamburg: United Digital Media. https://www.udg.de/en/udg-blog

Dr. Ralf E. Strauss is Managing Partner of Customer Excellence GmbH, President of the German Marketing Association, Professor of Digital Marketing & E-Business at Hamburg School of Business Administration, and Chairman of The CMO Community (www.cmocommunity.de), based in Hamburg/Germany. Previously he was the CMO EMEA Central at SAP, followed by being the Head of Product Management CRM Marketing at SAP, as well as he has been leading the digital transformation at Volkswagen Group.

Jonathan Becher is the Chief Digital Officer and Head of the SAP Digital business unit. Previously he was SAP's Chief Marketing Officer and has served as CEO for three different companies – NeoVista Software, Accrue Software and Pilot Software. He is a board member for the Churchill Club, Silicon Valley's premier business and technology forum, and Revel Systems, an iPad point of sale solution. He holds a master's degree in computer science from Duke University in North Carolina, and a bachelor's degree in computer engineering from the University of Virginia. You can follow Jonathan on Twitter: @jbecher.

Integrated Campaign Planning in a Programmatic World

Andy Stevens, Andreas Rau, and Matthew McIntyre

Of the many buzzwords and trends that float around the industry, 'Programmatic Advertising' may be the most prevalent and one of the most disruptive developments to media buying in the last 10 years.

The first question you might be asking about Programmatic Advertising is 'Why? What's so great about it and wrong with the people-based insertion order processes?' In truth, the approach to media planning still works in much the same way things used to be done. People are still (and will always) be key to planning a successful campaign. The value Programmatic Advertising brings to those planning is that it enables buyers to execute and deliver campaigns in ways that were just not possible before, right from the very start.

Automation and technology-driven algorithms support the planning set up and unlock efficiencies in execution that save time and enhance performance. The constant collection and updating of user data from campaigns and digital platforms, matched with real time optimization, allows planners to be agile and responsive in developing the plan throughout a campaign's lifecycle. Last and most crucially, Programmatic Advertising enables planners to continually refocus on the target audiences and campaign objectives, instead of proxies for your audience and KPIs, driving personalization and relevancy.

A. Stevens (✉)
Syzygy, Johnson Bldg, 77 Hatton Garden, London EC1N 8JS, UK
e-mail: a.stevens@syzygy.net

A. Rau
uniquedigital GmbH, Neuer Wall 10, Hamburg 20354, Germany
e-mail: a.rau@uniquedigital.de

M. McIntyre
uniquedigital UK, Johnson Bldg, 77 Hatton Garden, London EC1N 8JS, UK
e-mail: m.mcintyre@uniquedigital.co.uk

The following sections will outline the programmatic planning process in detail, and uncover what needs to be considered when planning a campaign that includes Programmatic Advertising.

1 Choosing a Framework for Programmatic Campaign Planning

For any media planning, it is essential to have a framework as guidance in building the best plan possible. It is key to understand the objectives of the campaign, who to actually talk to, what to say to the target audience and finally the desired result to achieve. Much of the heavy lifting needs to happen here, applying vigor and attention to detail to ensure the rational is sound.

For Programmatic Advertising, Avinash Kaushik's See-Think-Do (-Coddle) framework[1] works very well (Fig. 1). Kaushik's funnel based framework is user-centric and easy to understand and apply. This model helps to uncover the consideration stage of a user and their previous interaction with the brand, which in turn defines the chosen tactics and approach that are appropriate to the audience type.

Additionally, for Programmatic Advertising, when it comes to 'Think' and 'Do', the connectivity of data can play a much larger part in campaign strategy and requires additional focus.

This model will be referred back to throughout the planning and optimization process, and is used here as a starting point for the design of a brief.

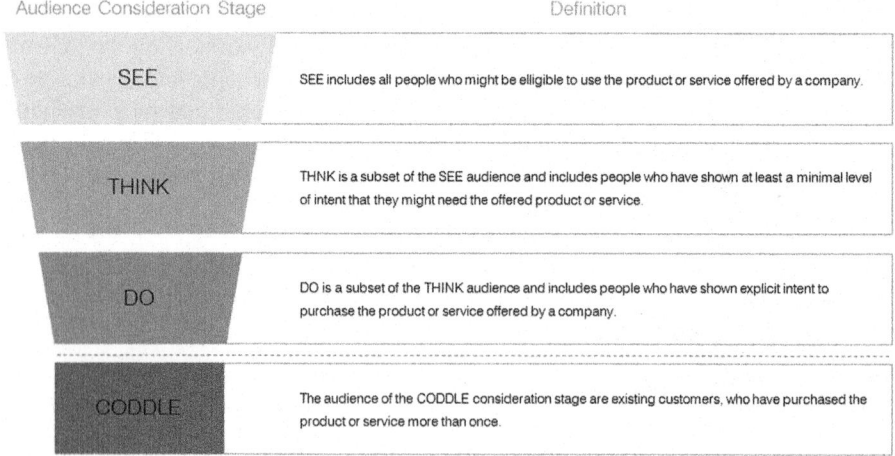

Fig. 1 SEE – THINK – DO framework by Avinash Kaushik

---

[1] http://www.kaushik.net/avinash/see-think-do-content-marketing-measurement-business-framework/.

2 Briefing

Before the actual planning can start, all necessary information needs to be at hand. Despite a comprehensive brief being key to planning a successful campaign, it is often overlooked by advertisers. These are the key elements which need to be considered before the planning can start:

2.1 Clearly Defined Advertising Objectives

The most important thing to know about a campaign is its aim. It is key to map campaign objectives against the user's consideration stage and then deduce the relevant KPIs. This will allow agencies and advertisers to track success for each objective and consideration stage. Finally, it is essential to define a singular core KPI for success. Others can be used for optimization and secondary goals, but every campaign needs a clear focus.

2.2 Well Defined Strategic Target Audiences

It almost goes without saying that there must be a clear understanding of who should be reached with your advertising, in order to deliver the right message. This doesn't just have to come from the marketing team. As well as market and audience research undertaken, business intelligence teams can have a significant role regarding audience segmentation and profile development on the right kind of customers.

2.3 Clear Picture of Existing Digital Platform Architecture

There is clearly no doubt about the importance of data and connectivity within Programmatic Advertising. With this in mind, knowledge about what systems are being utilized by other channels and stakeholders should be part of a briefing to map out how data can be connected or matched. This is not limited to media platforms such as ad servers demand side or bid management platforms, but should also consider CRM databases. First party customer data is extremely powerful if there is a solution worked out how to unlock it.

2.4 Wider Channel Mix and Strategy

For both brand and direct response/tactical campaigns it is important to build a picture of the marketing channel portfolio, as well as the topline strategy in place. This will provide a better appreciation of where and how display, video or mobile Programmatic Advertising should fit into the consumer's journey. Going back to

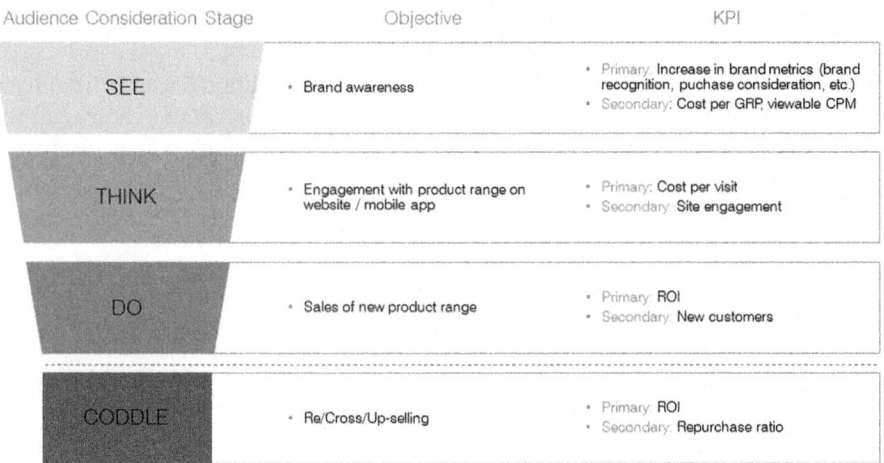

Fig. 2 Objectives and KPIs

the See-Think-Do framework, this can be applied at a multi-channel level to help guide the plan around the correct consideration stage.

> **Example**
>
> *To illustrate the different steps of programmatic campaign execution with its complexity and range of possibilities we will regularly add examples. In our examples we will refer to an imaginary multichannel retailer that just introduced a new product line for the coming season. The retailer sells its products through brick and mortar stores in various cities, an online shop and a shopping app. Additionally, the company holds CRM data from online and app customers as well as from a loyalty card program.*
>
> *The retailer wants to execute a full-funnel digital campaign to promote the new product line, ultimately driving sales. Besides information about audience, technical possibilities and marketing strategy, the essential element of briefing for success is the definition of clear objectives and KPIs, which are mapped against the according consideration stages (Fig. 2).*

3 Measurement

Now that a full and proper brief with comprehensive information about objectives of the campaign is available, the next question is around tracking and measuring the success of a campaign. Measurement must be clearly defined and agreed by all stakeholders upfront – Programmatic Advertising is no exception and should be measured like all other channels. Without clearly defined measurement and

Integrated Campaign Planning in a Programmatic World   197

attribution we won't know what results were achieved, what to optimize against and there will be very limited data to build learnings upon.

This should be done with care and precision, ideally presented back to everyone involved pre-launch to emphasize focus.

3.1   Measureable Actions

The first step is to do identify which actions or indicators match up to the defined KPIs. Site visits, specific URL visits (such as a conversion page), or button clicks are easily matched, but if the objective is not as clear then it is wise to find a close proxy or scope out a measurable indicator. An example would be time spent on site, as a proxy for high engagement.

3.2   Centralized and Integrated Tracking

With multiple digital channels and platforms running, many of the partners involved are going to offer their own slant on tracking and reporting with no appreciation of anyone else's activity. It is vital to get a centralized view on your marketing efforts, which ensures that conversions are not counted twice and that customer journeys are seen as one string across many channels.

3.3   Awareness of Potential Limitations

Though Programmatic Advertising initially focused on display, the growth of mobile and video Programmatic Advertising (and mobile video) is also accelerating, and this brings with it a whole host of new questions and issues which the industry is starting to address. The two big considerations right now are the different tracking currencies in mobile app and web inventory, Device IDs vs. cookies, and cross-device tracking, where we are attempting to build a single customer view that sees a customer across any device. Neither challenge is likely to go away, so awareness needs to be developed of how a campaign may be affected. It is best to evaluate any data available for suitability in addressing these issues, then to decide exactly what it will be possible to measure for any given campaign.

3.4   Attribution Models and Campaign Optimization

Once tracking has been centralized and all success indicators defined, a vital decision that can easily be forgotten about is how to accurately attribute credit back to the right media sources. The default position is 'last click' which carries well-known deficits and often may lead an advertiser to not fully appreciate all consideration stages of See-Think-Do correctly. Everyone should be experimenting

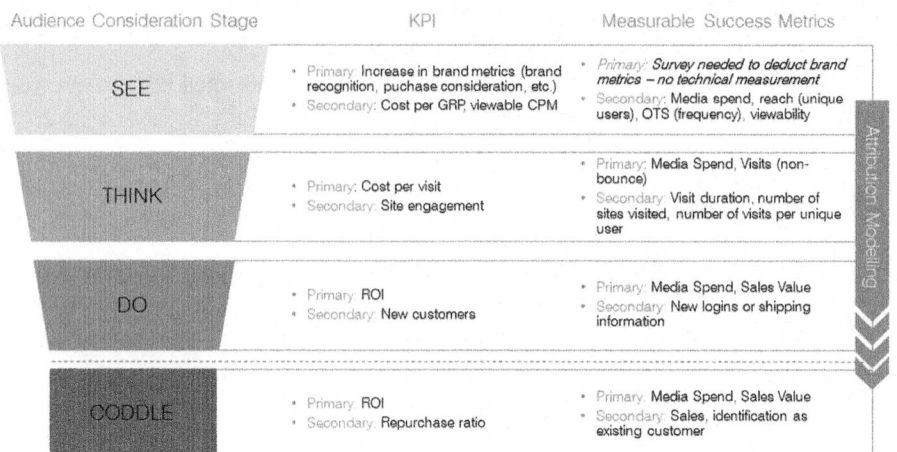

Fig. 3 KPIs and measurable success metrics

with alternative rules-based models at a minimum, potentially using different models to judge different consideration stages. To receive the best possible picture and prediction quality for all utilized channels we recommend dynamic attribution models that are built with advertisers actual cross-channel data. Attribution should be carried out across all trackable channels at once, to gain as accurate a picture of a user's journey as possible.

Example

The retailer uses a state of the art ad server and programmatic buying solution. Tag management & mobile tracking solutions are integrated on the online shop and shopping app. Tracking points are implemented to identify user behavior and to mark users for retargeting. Conversion tracking with tracking of sales values and deduplication of sales is set up. Due to logins and loyalty card information there are possibilities to track cross-device and offline sales. This enables the retailer to gather all relevant data points to report all technically measurable success metrics and according KPIs (Fig. 3):

4 Audience

The importance of the target audience was already mentioned. To get the most out of any investment it is necessary to find out who the right audiences are to reach. Audience discovery also means learning more about these segments to understand the best ways to engage with them.

4.1 Existing Customer Data Beats Assumptions

Existing customer data is gold dust. Not only is it a totally unique data-set to an advertiser, but it is the most powerful tool to discover key audience segments. It provides knowledge of the most valuable customers and if it is possible to connect this data straight into your programmatic buying platform it can be used to directly drive look-a-like activity. Relevant audiences can also be profiled manually using a Data Management Platform (DMP) to generate the insights, which will help in the development of programmatic tactics and strategies.

4.2 Site Visitors Tell Many Stories

Just like customer information, cookie data of any site visitor or Device ID of any app user can be utilized to profile it or use the user list directly for buying. Think about the different areas of a site or app and what each means. Successful tactics are to build specific messaging to suit visitors who have shown specific interests, or exclude audiences completely if their behavior suggests they don't fit the campaign audience brief.

4.3 Audience Profiles and Underlying Behaviors

Once there is a good idea of who the campaign audience is market research platforms such as Kantar TGI[2] or Experian Hitwise[3] can be leveraged to learn more about each segment. Research platforms are very helpful to identify traits such as consumer behavior, intent, marketing format receptiveness and preferences, device preference and consumption, socio-demographic factors, brand favorability etc. The tools and quality of data will differ by market so tool selection should be based on location of the campaign as well as the suitability of data collection methodology.

4.4 Low Hanging Fruit Within CRM Data

Everyone knows that retargeting is an extremely effective and above all efficient tactic for the bottom of the funnel, but it should also be explored how existing consumers can be leveraged to drive additional business. Tactics which may be suitable, dependent on campaign objectives, include up or cross selling to certain existing audience segments or driving increased advertising focus to users with a

---

[2] http://www.kantarmedia.co.uk/businesses/tgi/.
[3] http://www.experian.com/hitwise/.

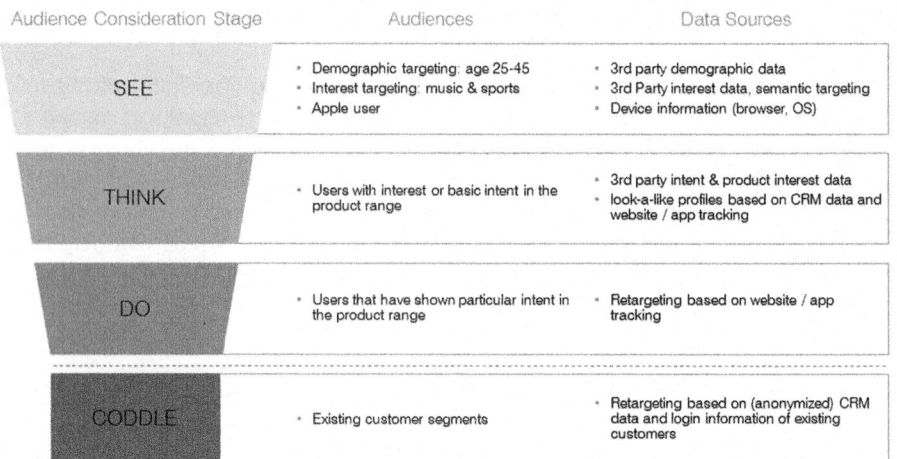

Fig. 4 Audiences and data sources

predicted higher life time value. The performance of these tactics will likely be among the best on plan for DO and CODDLE stage audiences.

> **Example**
>
> The retailer has customer information based on CRM data and market research that identifies the typical customer as 25–45 years old, living in a relationship, university degree, and particularly interested in music and sports content. The target group evenly uses desktop and mobile devices but skews particularly high towards apple products.
>
> The existing customers are segmented based on purchase history for potential re/cross/up selling as well as customer lifetime value to use as seed audience for statistical modeling (look-a-likes).
>
> There is granular tracking (pixel/SDK) implemented on the online shop and shopping app to identify users who recently were looking for products and/or had a purchase. We can use these for retargeting or look-a-like modeling. To avoid stalking all users who had a purchase during last 30 days are excluded (Fig. 4).

5 Campaign Strategy and Set Up

Planning and actually setting up programmatic campaigns are two tasks which are much closer together than they are for IO-based media buying. Instead of re-packaging everything an advertiser or agency has learnt so far and speculatively requesting proposals from a select group of media owners, demand parties have to look at the platforms themselves and match all their data and specific audiences to targeting rules that are available to them. One of the scariest parts of doing programmatic advertising is the sheer amount of targeting options and combinations available to every advertiser. No algorithm will conquer the deep sea of options available without expert help or theoretically an unlimited time and budget to run simulations. Here are just a few of important aspects for setting up a campaign.

5.1 Inventory Sources and Buying Technologies

Every campaign is different. Every inventory source and according programmatic buying technologies are different as well. When making a choice of a programmatic buying tool, the first thing to look at is what type of ad inventory is part of the campaign? Some inventory sources have a closed ecosystem where you have to use their own or very specialised buying technologies to access their media; a really good case of this is in the paid social space und various forms of marketing partners. After that the next considerations are about a variety of areas, which will define the strengths needed for your buying technology. What possibilities exist to connect 1st party data to the platform? Mobile, desktop or both? What formats or size of creative is relevant for the campaign? Is there a focus on mobile apps? At which market is the campaign directed? All these are important questions that inform the choice of inventory sources and buying technologies.

5.2 What are the Priorities: Performance or Premium

The structure of programmatic buying has developed as the industry has matured to include multiple routes to market for ad inventory (Fig. 5). This has naturally created a hierarchy where publishers are looking to both maximize their valuable

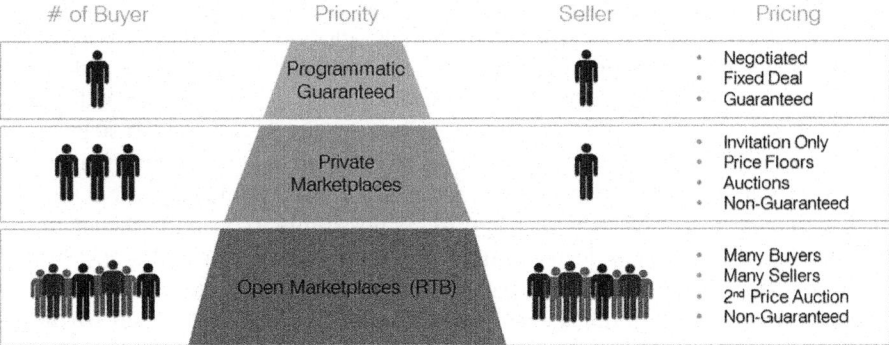

Fig. 5 Programmatic buying hierarchy

premium space whilst ensuring they sell all of their available ad space. Open exchanges are the original form of Real Time Bidding (RTB), where many buyers and many sellers engage to transact ad inventory using a simple auction style model, during the milliseconds after a page starts loading to the user. RTB through the open exchanges is still where the bulk of performance campaigns are delivered – but private buys can also be useful and are increasingly worth testing. Private deals come in two forms; Private Marketplaces (PMPs) or Private Auctions operate in a similar way to RTB, but may have higher minimum bids and will have only a select amount of competitors. Guaranteed (or sometimes known as Direct) deals are like IO buys, with a fixed rate and volume agreed – but combined with the advantages of Programmatic Advertising such as rejecting specific users to don't waste money on. In some markets where publishers have been slower to adapt to an open exchange model and it may be required to deliver a larger portion of the campaign inventory through private deals.

5.3   1st Party Data is the 1st Choice for Targeting

As mentioned very clearly throughout this article, 1st party data is the gold mine that any advertiser has exclusive access to. All the audiences that have been previously uncovered and set up during the audience discovery phase can be part of a target audience. These will be usually the best performing targeting lines. To add scale to 1st party data look-a-like strategies can be developed to algorithmically identify similar users to the top performing audiences. The other important use of 1st party data is to exclude any audiences that an advertiser definitely does not want to message. If a campaign is for new customer acquisition, it is recommended to block all of the existing customers to minimize wastage.

5.4   Addition of Targeting Features to Scale Prospecting

The extra features available will depend on the selection of inventory sources and buying platforms. There are many different tactics that can be tested depending on campaign objectives. Contextual keyword data as well as site category and domain targeting are useful options for objectives closer to the Think and Do stage of the funnel. 3rd party audience data, from suppliers relevant to your region and industry, can be extremely powerful for the See and Think consideration stages, but may not be focused enough to drive sale objectives. A possibility to push the boundaries, is to integrate live API data feeds straight into the platform that will adjust bids or activate specific tactics at key moments in time. Possibilities are widespread, for instance advertising umbrellas when it starts to rain or second screen advertising on mobile devices when a TV spot hits the air.

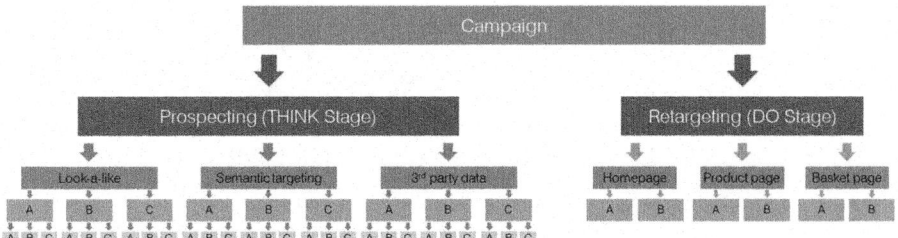

Fig. 6 Sub-segmentation in a pyramid structure

5.5 Campaign Structure Makes a Big Difference

Defining campaign structure during initial setup is an important first step to allowing proper analysis and continuous optimization. A pyramid style structure (Fig. 6), breaking down the audiences in different sub-segments, is a helpful tool to lay out a campaign structure. This enables an advertiser to build the priorities within the rules and settings to properly match against campaign objectives; leaving the buyer with complete control to point the algorithms in the right direction.

5.6 Recency and Frequency

There are few things more annoying to consumers online than being stalked by retargeting campaigns. It is not just annoying for consumers, most often it is also inefficient for advertisers. So the simple things should be kept in mind, resulting in a clear decision how to control recency and frequency within a programmatic campaign. Recency of a tracked behavior is also a way to control data quality, as targeting users with stale targeting information will both be ineffective and user-unfriendly.

5.7 Controls and Media Quality

Wherever there is money to be made, criminals usually will try and find ways to exploit the system – the programmatic markets are no different to many others in this respect. It is imperative to put technology and manual checks in place to avoid ad fraud in its many forms. Partnering with one of the brand safety specialist firms is a technological way to do this. Other effective possibilities are to use Private Market Places PMPs, avoiding non-transparent inventory sources and buying with whitelists (approved sites) and blacklists (disapproved sites). Even when the ad space is real, content categories and viewability are indicators of quality that you should be monitoring at a minimum.

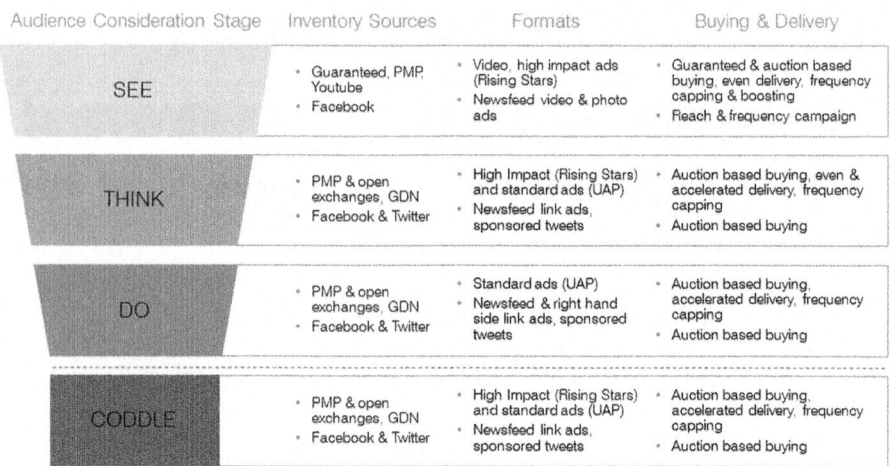

Fig. 7 Inventory sources, formats, buying and delivery

> **Example**
>
> *The retailer uses different inventory sources and formats as well as buying and delivery methods. The selection is based on campaign objectives, audience profiles and data availability (Fig. 7).*

6 Creative Set Up

Although media is a central point of Programmatic Advertising, finding the right media is only half the job. If the creative and messaging isn't right, then all of the advantages and efficiencies can be diminished easily by irrelevant creative. With all the advances of programmatic buying for media, it would be a shame not to share some of that with the experts in the creative departments.

6.1 Granular Campaign Structure and Matching Creative

When a campaign structure is split out according to tactics, it needs to be assured that the best creative is assigned to each set. For keyword or audience data – an option is to group them by product or message. For 1st party data it can get even more specific as the knowledge about a user becomes more concrete.

6.2 Targeting Data Informs Personalized Creative

Dynamic creatives have been around a long time for retail retargeting – but that is not the only way they can be used. Ideas for available targeting possibilities should be shared with the creative team to develop new ideas for triggers using 1st and 3rd party data. Possibilities are endless: time bound messaging to match a flash sale, hyper-localized messaging, real-time quotes in ads, different backgrounds to match the weather. By pairing a dynamic creative with a targeting set that uses the same dynamic data feed, a lot of 'default ad' wastage can be eliminated.

6.3 Sequencing and Storytelling

Recency and frequency are equally important for creative setup as for media controls. Setup can include tracking of a user along all campaign touch points and sequencing of creative based on user response. This is an easy tactic to keep users engaged and eventually find the most effective message. To guide users down the funnel thoughts should be about storytelling and what message a user should receive at what consideration stage.

Example

The retailer uses an overall creative idea to deliver a consistent brand experience. Within the overall creative idea there are different assets and messages for the different target groups to ensure relevant communication. To reduce creative development and setup costs, dynamic creative templates are used to show relevant visuals or particularly in case of retargeting the most relevant product. Additionally different triggers will be used to dynamically adjust the creative, for instance:

- Products, services or product categories based on site behaviour
- Time based triggers to layer in countdowns or sale periods
- Availability ("only x items left")
- Vouchers for existing customers/new customers
- Weather based offers ("rainy Sunday special")

7 Landing Experience

Another element that can ruin a great programmatic campaign is a bad landing page. When a user lands on an app or website, the experience should be just like a smooth arrival at an airport. No one wants a bumpy landing and definitely no one wants to find out that the luggage was left at the other end. These are a few of the parameters that help define a smooth landing experience:

- The consideration phase of the user
- The type of media and the messaging they have engaged with
- The device they are on
- Any information on previous site visits or transactions for the user

Users can receive visuals they are familiar with from the ads they saw before they came to the site. Based on additional information of the customer about demographics, geolocation, ad reception and onsite behavior, the optimal landing page can be assembled dynamically.

The content of the landing page will be dependent on all of these variables, but it is also important to remember the key messages behind the campaign or any proof points that you need to deliver. Unique Selling Points (USP) or other influential information should be woven through in the most suitable format.

> **Example**
>
> *The retailer extends the consistent communication from creative onto landing pages. The different types of interaction with ads, the online shop and the app are tracked (view, click, video completion, abandoned shopping basket etc.) and stored together with target group information and the creative execution the user received. According to the consideration stage and tracked ad or site interaction, the user will land on a dynamically adjusted page that will pick up the visuals or products of the creative and additionally show specific information or filters based on consideration stage or target group (Fig. 8):*

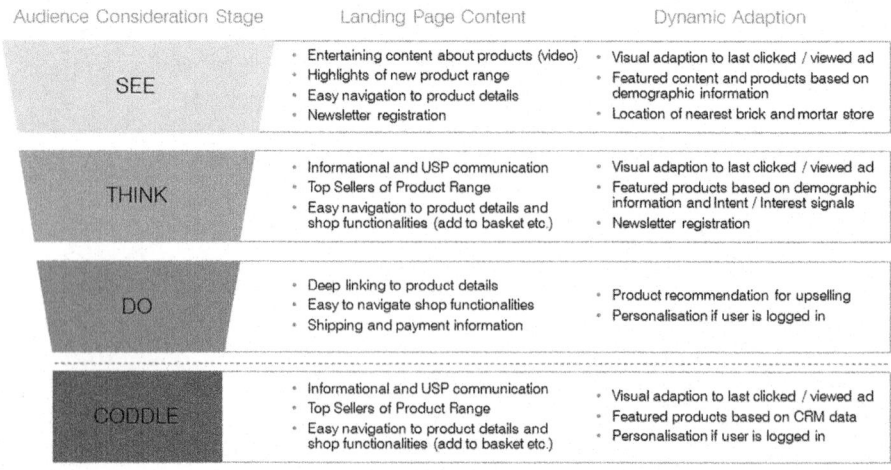

Fig. 8 Landing experience

8 Campaign Management and Optimization

One of the big advantages of programmatic, that was mentioned at the outset of this piece, is the ability to make continuous optimization and changes to campaigns. The actual management of a campaign is about finding the right balance between science and art. Machine optimization is very good at making small continuous tweaks, but humans are needed to set the strategy and then make broad and divisive changes based on both qualitative and quantitative data.

8.1 Sufficient Data Is Important for Informed Campaign Optimization

To get the best results out of the algorithms and machine learning features in programmatic platforms, significant volumes of data have to be generated. In some cases it may make sense to rely on secondary KPIs that deliver larger volumes of optimization data, to accelerate the process from test to learn. Once significant data volume is gathered, an advertiser can start to turn on the optimization algorithms which will make changes to every single bid request that matches the campaign setup. This is done in real-time, based on 100's of factors, and should drive automatic improvements to your campaigns.

8.2 As Campaigns Develop, Humans Need to Guide The Machines

Machine optimization engines are great at making rational and logical decisions based on all the data they have available to them – but they can't make decisions they haven't been trained for. Machines can't create new tests, won't appreciate trends or events happening in real life or know that there is a key seasonal date approaching unless you tell them. Analysts are needed to look at the larger trends and make the changes, which will maximize the power of the machines for you. The responsible (external) organization has to be properly staffed and structured to manage this ongoing component.

8.3 Continuous Optimization Drives New Campaign Briefs

As campaigns develop and grow, learnings should be constantly made and recorded. These can feed straight back into the campaign to improve every aspect of campaigns as they are still going. Every aspect of programmatic planning feeds into this optimization cycle, which drives continuous performance gains (Fig. 9).

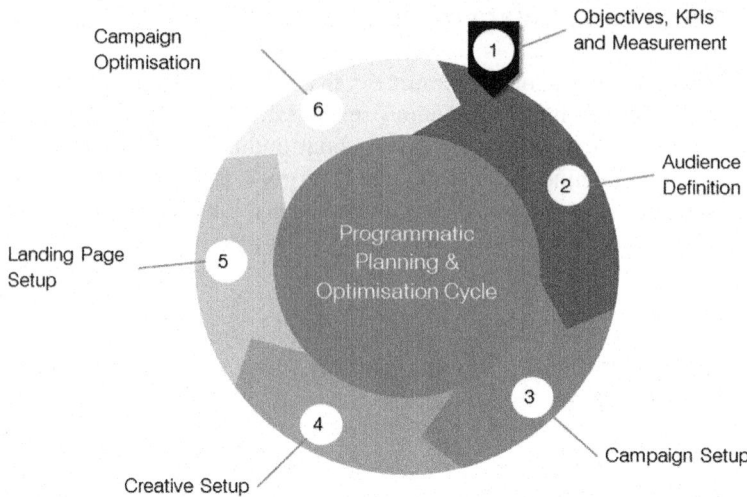

Fig. 9 Programmatic planning and optimisation cycle

9 Conclusion

In many respects, planning a programmatic campaign is very similar to any other media planning activity, but it does have its nuances. These nuances require a very different set of skills and expertise compared to IO based buying. Technological knowledge, an analytical mindset and user centric thinking must be much more advanced. There is a much heavier emphasis placed on data and the connectivity of buying platforms. A significant advantage is the ability to continually optimize throughout a campaign, utilizing machine learning and advanced algorithms to drive increased performance against your set KPIs.

That's not to say that there are no problems which need to be faced. Cross-device tracking is high on everyone's awareness radar, but there are multiple approaches currently in the market and the industry has not yet started developing standards for vendors to follow. User logins across multiple platforms are the best way to map devices together but current offerings are being controlled by a few large publishers who have enough scale to make it viable. This is a very powerful approach, but marketers do need to be wary of potentially getting tied to certain inventory sources. While not as accurate as logins, more and more probabilistic solutions are being developed. These are definitely helping to guide advertisers in the right direction and the accuracy of these methods will undoubtedly improve over time.

Issues of inventory differ by market. Markets at an earlier stage are dealing with scalability problems while the developed markets are now facing questions about quality and fraud. In all situations, we are now seeing the industry start to come together and fight for standards and solutions. We are starting to see partners in all capacities make a concerted effort to legitimize the industry.

Potentially the biggest issue programmatic advertising is facing is actually one that the whole digital industry is facing together – data. Consumers are becoming more aware of personal data and privacy regulations. Lawmakers are sometimes working with rules and policies that are still not quite fit for purpose and strict local data protection philosophies can occasionally close doors, which are open to many others.

Even taking into account all of these factors, there are incredible opportunities for marketers within Programmatic Advertising. The combination of brain, data and heart, necessary to execute successfully across all consideration stages, is likely to be within your teams already. Armed with a solid planning framework, the right combination of people and machines and the advice above – advertisers and agencies will be perfectly placed to deliver innovative and successful programmatic campaigns.

Andy Stevens As the founder of uniquedigital and the current Chief Operations Officer for the parent company, Syzygy AG, Andy has over 20 years experience of delivering digital marketing campaigns across the globe. For the likes of HP, NatWest, Vodafone, Barclays, Mazda and Avis Andy has consistently looked for innovative solutions and has been an early adopter of some of the most influential media technologies brought to market. His passion for delivering success fuels the need to consistently push the boundaries of digital opportunity

Andreas Rau As Director Paid Media and member of the board at uniquedigital Hamburg, Andreas is responsible for strategy and development of innovative digital media campaigns across multiple channels. Being a proven expert for data driven advertising his main focus after joining uniquedigital Hamburg was to develop and establish the internal programmatic and social advertising unit. Andreas looks back on well over ten years of international digital marketing experience. Within senior and leading positions at Yahoo! Europe and United Internet Media AG, Andreas delivered successful digital campaigns for the likes of Zalando, Deutsche Telekom, Otto Group, Commerzbank, Comdirect or Ebay and was responsible for the implementation of digital marketing innovations, processes and technologies for data driven and programmatic advertising, targeting and yield management.

Matthew McIntyre has worked at uniquedigital London since 2011, looking after media planning and operations for clients across a wide variety of verticals. Currently, as Programmatic Associate Director, he manages the strategy and integration of programmatic media for all clients and leads the internal media trading team.

Evolution of Digital Campaign Design and Management

Nils Hachen and Stefan Bardega

Digital campaigns have always been a hare and tortoise game. Looking back, this common thread runs through the past 20 years of digital advertising. Who is in charge? Is it the creative agencies or the media agencies? Who has the best technology, and which technology should be implemented? Which hype will become a real trend? These discussions have been going on for 20 years and will continue well into the future.

1994 was the dawn of digital advertising and thus also of digital campaign design. What was relatively simple in the beginning (because there was only a single advertising medium) has grown to become a herculean task. Once agencies and customers had developed a sufficient understanding of the display business and were able to implement it, the success of search engines at the end of the 90s gave rise to performance marketing agencies, along with new providers and approaches which had to be learned all over again. A second branch of digital advertising developed relatively rapidly alongside conventional display advertising. The first (specialised) agencies established themselves between the late 90ies' and the early'00 years. They developed concepts for search engine marketing, used digital distribution partners – the affiliates – and created advertising networks, which produced further evolution and an at least temporary dichotomy of the digital advertising landscape (see www.marketing-boerse.de).

The main difficulty lay in how to classify the new medium and integrate it into the familiar structures. Silo mentality instead of 360° or even better 365 – individual test campaigns instead of campaigns from a common mould. A comparable

---

N. Hachen (✉)
Zenithmedia GmbH, Louis-Pasteur-Platz 3, 40211 Duesseldorf, Germany
e-mail: nils.hachen@zenithmedia.de

S. Bardega
ZenithOptimedia, 24 Percy Street, London W1T 2BS, UK
e-mail: stefan.bardega@zenithoptimedia.com

situation was also apparent during the last few years in the field of social media, where the mistakes of the past have been repeated.

A general problem of the digital channel was and is that it is very much technology-driven. Discussions revolve around technologies instead of solutions, and often at a level which tends to deter rather than inspire those market participants who don't deal with the subject matter on a daily basis. The options with regard to dialogue, response or (trans)action were initially rather limited. Digital advertising and digital advertising design were based excessively on the offline world and also on print.

While the content was present in principle, it was offered in a one-dimensional way and didn't have the coverage of established offline channels. However, the content providers needed this coverage to justify the use of the internet. In order to attract as many users as possible to websites, the content was provided free of charge – a problem the industry is still struggling with today. Users were willing to pay for access, but they wanted the content to be free. It was necessary to find and secure a way to finance the content; or even better to generate another source of income. The obvious plan was to transfer print advertising methods directly onto the web – and this marked the dawn of digital advertising.

The very first campaign was a complete success. In 1994 the telecommunications provider AT&T launched a first trial with a full-size banner on the Hotwired website. The unexpectedly high click rate of 40 % exceeded the expectations of the visionaries. The message at the time was relatively simple: Click here. A new genre was born before most people even knew how they were supposed to use the internet (see Heise.de and cpc-consulting.net).

This pilot project triggered a certain amount of euphoria. Finally one could earn money with the new channel. Spurred by the initial successes, new advertising methods were developed on a regular basis. A few of them disappeared again, while others have their place to this day. There were no technical standards. Every marketer developed their own system, ad servers didn't exist yet and there was no clear solution for a creative approach to these tiny formats. The result of this was that a supposedly simple banner had to be "created" for every individual marketer. The technical specifications were too different. This resulted in the first negative reactions to the new channel, which, at this time at least, was anything but transparent. Organisations and agencies worked with the marketers to develop definitions, but these were only established years later. Online advertising becomes incredibly complicated. Not only is the confusing terminology unclear and impossible to understand by non-technical marketing and media people, but the technical implementation of campaigns also requires considerable know-how.

Meanwhile, we have to live with the fact that around 90 % (figures differ country by country) of all internet users don't click on the advertising even though they see it. Both measuring and control aspects are becoming more important. Marking the generated website visitors makes them visible. More can be learned about the target group via targeting, and the acquisition of data makes it possible to form statistical twins, giving a much more accurate picture of the clicking target group. Increasing interaction in the advertising media itself plays a large role, as does creation – at

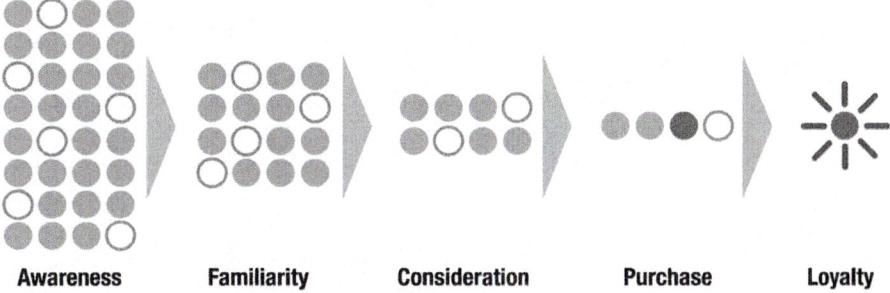

Fig. 1 The traditional McKinsey buying funnel (Source: www.mckinseyquarterly.com)

least in the sense of a "call to action". Accessing the website via the advertising media is no longer the sole objective.

1 Fundamental Campaign Objectives

The objective of all campaigns is essentially to elicit something in a (potential) customer. At least a "mental click" – better yet an action. This could be a request for further information, a download, a test drive, a booking or an order. In light of budgets, the following questions come up as well: Which channel should I use with what objectives and what budget do I have to make available in order to achieve an optimal (digital) channel mix. At this point the customer comes back into play. Who is the target group? What does the target group do? What does the advertiser have to put on the table in order to gain potential customers? In which phase of the decision-making process is the customer, and how can I bring the customer from this phase to the next one? (Fig. 1).

The objective of strategic campaign planning is to obtain the best possible mix of branding and performance components. How much brand do I actually need in order to generate the best possible performance right from the start, and how much performance can I actually achieve with my brand?

2 Traditional Advertising Impact Analysis Was the Basis

The fundamental task of an analysis is to process all data and facts from a campaign and to interpret them accordingly. That's a tall order – things often look quite different in reality. In some cases no measurement is taking place at all, and if data is generated it often takes the form of Excel spread sheets which are filed neatly in some folder (physically or digitally) and left to gather dust. Companies that are a bit more advanced do at least obtain transaction data, but more often than not still only consider these occasionally. The offsite and onsite generated data are rarely linked, and/or there is no standard system of measurement. There are only a very few companies today who are in a position to truly understand what effect an individual measure has had on the specified key performance indicators (KPIs).

The goal of an advertising impact analysis is to determine how effective and/or efficient a campaign has been. Naturally, the individual instruments are first considered individually and optimised, in order to subsequently put them into context and integrate them into an overall view. Among other things, this method has made it possible to illustrate the correlation between TV advertising and online searches.

A large amount of data is traditionally collected in the digital media: Ad impressions, views, clicks, visits, unique visitors, fans, click through rate, interaction rate, engagement rate, downloads, sales, length of stay and many more. This documents the effect, but not the cause. Add to this the fact that another aspect hasn't been considered at all yet – what do these numbers mean in relation to one another? Is there a benchmark? And if so, how do my numbers compare?

How did the attitude of my users change over the duration of the campaign? Did the campaign help to increase awareness? What changes occurred with regard to preference and first choice, or recommendation? By analysing these numbers, I can also measure the effects that are relevant for bricks-and-mortar retail and/or POS.

These numbers are typically determined by way of surveys, either automated via a questionnaire within a campaign or on the corresponding website. Panels are interviewed in person for more extensive or longer-term campaigns. By comparing this information with the media data and, if applicable, also including CRM data, one can obtain a truly informative picture.

The problem that arises here is that this is purely an ex-post analysis. The advertiser can only determine whether or not his campaign was successful after the event. In the best case scenario, it is possible to determine which advertising material moved the user to act, and it is extremely useful to be able to confirm and, where applicable, correct the target group. However, these surveys involve considerable effort and expense, making ongoing analysis even more difficult.

Both the customers and the agencies want reliable forecasts that can model potential future scenarios. Past experience is often used as a basis and supplemented with the current campaign types. Initial digital approaches (see Adobe.com) exist already. They use predictive modelling, for example to simulate paid search scenarios. The goal must be to implement this at a minimum for all digital campaigns and to offer a solution for all media channels over the medium term (Fig. 2).

3 Big Data Becomes Relevant Data

Every step and every movement of the user has been documented for years now in order to optimise campaigns on the basis of this information. The data isn't complete in most cases, and important information is missing: Either not all pages are pixelated, the content or value of the shopping cart is unknown, or the potential customer's decision journey is insufficiently documented. In addition, this

Evolution of Digital Campaign Design and Management

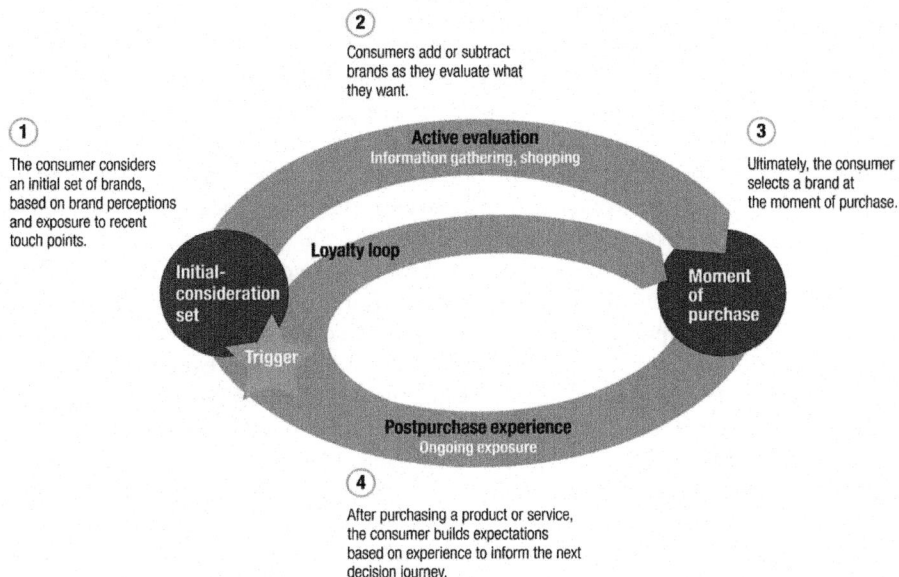

Fig. 2 The consumer decision journey by McKinsey (Source: www.mckinseyquarterly.com)

information is collected and processed by different systems, measurement differences arise and the data cannot be compared at a glance. Against this background, proving and illustrating the advertising effect is a high priority task.

In this regard, Big Data is often confused with the pure collection of data. But it's not just about tapping into as many sources of data as possible, but rather about understanding the data, generating added value and thereby achieving a competitive advantage through better information. The better someone uses the data, the better the understanding of the respective customers, the more precisely articulated messages, the more efficiently the inventory buying. And all of this on the basis of my formulated objectives.

Let's be honest: Nobody cares about ad impressions or click rates. The advertiser either wants to generate sales, or at least lay the groundwork for a sale or improve his key brand KPIs such as brand awareness, preference/top of mind or buying intention.

This requires tools, which provide reliable data as well as employees or agencies who can interpret it.

4 Data as the Basis for Programmatic Advertising

Declining click rates have been countered with a continuous stream of new technologies and formats. The measureable success was mostly short-lived, leading to the next advancement. This is comparable with a Formula 1 team, which develops the car for the next racing season while the current season is still underway. The situation was aggravated by declining profits in the classic display business. Unlike digital video which is still a scarce product, it wasn't possible to monetise the range of "classic" digital advertising formats sufficiently. The portals with broad coverage provided residual space inventory to the so-called performance networks, and these third-party providers marketed these spaces. Performance networks or blind networks had the disadvantage of lacking transparency. Advertisers had no overview of where their advertising was actually displayed.

If one ignores the "disadvantages", the campaigns were very successful. Add to this the constant advances in data utilisation. Initially things were optimised for clicks – on the assumption that places where people click a lot also contain the right target group. The results could then be optimised after the event. Which users actually performed the desired action? Which pages triggered the conversion? Both sides were under pressure. Due to the hybrid monetisation, the marketers were only paid if the defined action took place. This meant that a "poor" creation without corresponding call to action frequently took place but didn't generate any sales, which were defined as the basis for assessment. The advertiser still received ad impressions and thus a coverage which they received more or less free of charge. The marketers looked for other ways to optimise their sales.

5 Why Programmatic Advertising?

The quality of a contact, combined with the ability to reach the right target group at the right time and the right place with the right message, gave rise to agency trading desks. Automation – of purchasing, processes and budget management. All summarised under the terms 'programmatic advertising'. 'realtime performance' or 'realtime advertising' (see BVDW (Hrsg.): Realtime Advertising Kompass 2015/2016). The precursor, realtime bidding, was considered a pure performance measure and denounced as simply residual space marketing for a long time. Today, new methods pursued through programmatic advertising are also of great interest to premium branding customers and thus used in such segments as well.

While it was initially almost solely a matter of generating sales, or at least registrations, the new programmatic advertising approach now satisfies both demands of the digital world: programmatic advertising provides performance not only with regard to the required user action, but also offers the ability to achieve branding and awareness effects for brands and products.

With the current data and technologies, it is now possible to access not just target groups but target persons. With more information about the status of the respective buyer decision process of the individual user, one is in a position to define the price for the contact even more precisely and to transmit the correct message. The user of a stationary computer during working hours has a different reception pattern than a user of mobile end user devices during the evening.

The objective or final stage of programmatic advertising should be to take further development of today's standards to the extreme. In no way is the goal to stalk the users and assault them with the same advertising material time and again. It's no accident that cookies are deleted on a regular basis these days. After all, the goal of the advertiser and thus also the agency has to be to send different messages which are matched to the buying decision making process of the potential customer. It is crucial not to create a sense of being pursued. We have to place the messages so skilfully that users don't even realise that we have certain information about them and their online behaviour.

Innovations always have to struggle with challenges or restrictions during the early part of the development process. The advertising spaces were more or less attractive, and the advertising formats one could book were standard AdBundles with small formats. In order to actually achieve branding effects in addition to performance, advertisers and agencies required the right pages (attractive and with broad coverage) and large-format advertising material. Jamming an expensive vehicle into a skyscraper makes no sense.

6 Programmatic Advertising as Competitive Advantage

The agencies use so-called private exchanges to secure premium pages, which guarantee the coverage required by the advertisers. In addition, large-format advertising material is used more often and leverages the advertising effect. In order to verify the effectiveness with regard to brand values, the agencies perform accompanying market research. First an online reference measurement is performed before the start of the campaign. Cookie tracking is then used during the campaign to determine which panel members came into contact with the advertising material. These members are surveyed again using an actual test measurement after the end of the campaign.

Figure 3 shows the results of a study in which large-format, image-forming advertising material was used exclusively in high quality environments provided by the Vivaki agency in the form of private marketplaces. The Vivaki study proves the success of programmatic advertising with regard to the advertising effect.

"The results of the campaign provided proof that automated buying has rightly established itself as an integral component of branding campaigns", explained Marion Kölling, Head of Audience on Demand (see Adzine.de) (Fig. 4).

7 Conclusion: Theses and Visions

(1) **Data are gaining even greater importance.**
It's not about the quantity, but rather about the quality and interpretation of the data. The vision calls for a sort of cockpit which answers questions in a supported manner. How much will performance change if less is spent on the brand? How much budget spending is required in order to achieve a specific performance or specific brand values?

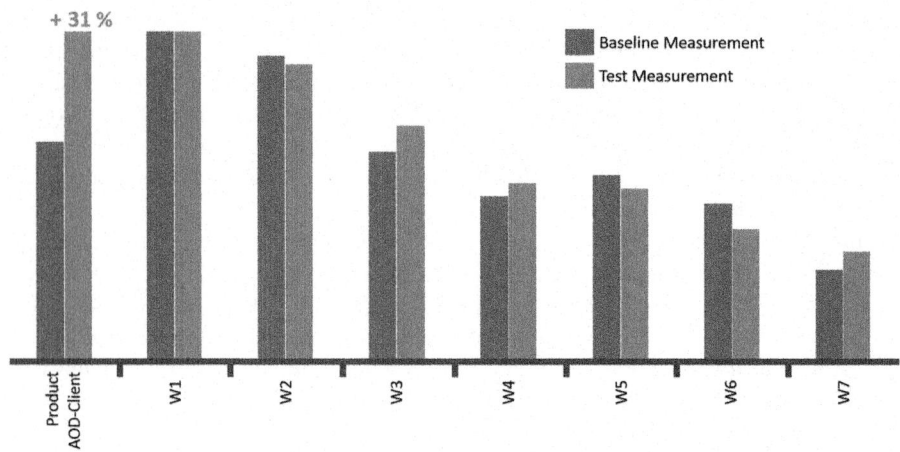

Fig. 3 Brand awareness in the competitor comparison (Source: Graphic from Vivaki)

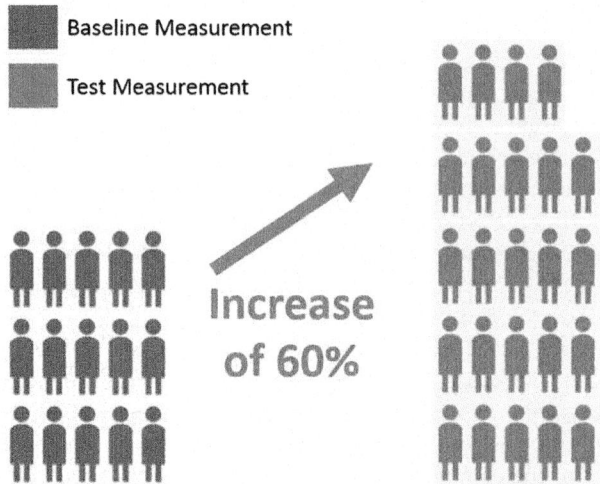

Fig. 4 Success of programmatic advertising at product level (Source: Graphic from Vivaki)

(2) **Online and offline data are merging.**
The digital advertising industry needs reliable "currencies" for the optimisation of target groups and the output of advertising material. In the future, it is not only digital pure players who will be involved, but rather all those who bring together reliable data which (could) play a role in optimising a campaign. Whether we will still be talking about cookies then or actually using digital fingerprinting is anyone's guess.

(3) **Buying models will change.**
In the future, customers will want to purchase the fulfilment of their objectives, and the agencies will have to allow themselves to be assessed and compensated by this measure. Awareness or purchase intention points will potentially be

purchased at a fixed "cost per point" in the near future, increasing the risk, which agencies will (have to) take.
(4) **Messaging is becoming even more important**
It is not only about the right audience, at the right time and place for a valuable price any more. The industry is moving from one message fits all to a consumer centric approach. Customized storytelling along the decision path becomes more important and is one of the new challenges for advertisers and agencies. Relevant data and the capabilities of the programmatic approach delivers campaigns that are more efficient and more effective with a win win for both sides – for the advertiser and the potential customer.

A realistic vision is the transfer of personalised/profile-based advertising from digital channels to, for example, TV. It can be assumed that the technical ability to provide every user with individually compiled commercial breaks will be available in the foreseeable future. Players from across the media landscape – advertisers, data management platforms, sell-side platforms and, slowly, TV broadcasters and networks – are starting to come together to drive innovation toward making programmatic TV a reality (see theguardian.com "TV ad buying will soon be fully automated"). History is repeating itself: Display → AdNetworks → Realtime Advertising → TVC → DRTV → Programmatic TV → Realtime Advertising in TV.

Bibliography

Adobe.com. http://www.adobe.com/de/products/adlens/adlens-for-search.html. Accessed 15 Oct 2013.
Adzine.de. http://www.adzine.de/de/site/artikel/10090/adtrading-rtb/2014/04/vivaki-programmatic-buying-funktioniert-auch-fuer-branding. Accessed 25 Apr 2014.
BVDW (Hrsg.). (2013). *Realtime advertising Kompass 2013/2014* (pp. 74–75). Düsseldorf: BVDW. CPC-Consulting.net. 2014. http://www.cpc-consulting.net/10-Jahre-Online-Werbung. Accessed 25 Apr 2014.
Heise.de. http://www.heise.de/newsticker/meldung/Ausstellung-ueber-10-Jahre-Online-Werbung-109909.html. Accessed 25 Apr 2014.
Marketing Börse. http://www.marketing-boerse.de/Marktuebersicht/details/Performance/10846. Accessed 15 Oct 2013.
Orlet, P. Berlinbuzzdotnet. http://berlinbuzzdotnet.files.wordpress.com/2011/07/mckinsey- quarterly-com1.png. Accessed 28 Apr 2014.
Orlet, P. Berlinbuzzdotnet. http://berlinbuzzdotnet.files.wordpress.com/2011/07/mckinsey- quarterly-com-21.png. Accessed 28 Apr 2014.
Springer Gabler Verlag (Hrsg.). *Gabler Wirtschaftslexikon*. Wiesbaden: Springer Gabler. Keyword: Werbewirkung. http://wirtschaftslexikon.gabler.de/Archiv/57379/werbewirkung-v7.html. Accessed 25 Apr 2014.
Theguardian.com. http://www.theguardian.com/media-network/media-network-blog/2014/sep/11/programmatic-ads-targeted-tv-audience. Accessed 12 June 2015.

Nils Hachen is Chief Marketing Officer Digital at ZenithOptimedia and the Head of Digital Consulting. In this position, he supports well-known customers from the finance, telecommunications, automotive, FMCG, health, luxury and tourism industries. Before that, Nils Hachen was Executive Managing Director at Zenith and Performics and Media Director at denkwerk.

Nils Hachen has lectured for many years (including at DHBW Mosach, BA Dresden, DDA Düsseldorf, MAK Cologne) and headed the Fachgruppe Performance Marketing at the BVDW from 2009 to 2013.

Stefan Bardega spearheads ZenithOptimedia's digital strategy and global capabilities. As Chief Digital Officer, Stefan sets the vision for how digital elevates ZenithOptimedia's Live ROI proposition and he is responsible for driving digital innovation around the network. Prior to joining ZenithOptimedia, Stefan was Managing Partner at MediaCom, heading up digital in the UK across all platforms. In 2009, Stefan launched MediaCom Beyond Advertising, the network's content division, which has now rolled out around the world. Before taking charge of this launch, Stefan was Digital Strategy Director at MediaCom UK. Stefan has also worked at All Response Media, where he was Head of Digital, and at Outrider, where he was Digital Research Manager.

Realtime Data Accelerates Online Marketing

Kolja Brosche and Arun Kumar

The way how digital media can be bought is changing fundamentally, from direct buy to automated buying, combined with realtime data. This evolution is not just a trend, it is a paradigm change. This change is driven by a massive development of technology on the one hand, and the need to optimize processes and the use of tools on the demand side on the other hand. Big agency networks are proclaiming the move to automated media buying in realtime since a few years. 50 % of all media buys will be automated until end of 2015, regardless of the media channel, stated Matt Seiler, former CEO of IPG Mediabrands (O'Leary, N. (2014). Matt Seiler is out to remake and automate the media agency world (Online). http://www.adweek.com/news/advertising-branding/matt-seiler-out-remake-and-automate-media-agency-world-155047. Accessed 26 Jan 2014). The combination of ad space with audience data for more target-oriented and more affine media buys itself is not revolutionary. The resulting possibilities for online-marketing, with the granularity of the data, the miscellaneous technological capabilities and especially the advantage of the data availability in realtime, geo-focused and multichannel, are tremendous.

Will this data bring more efficiency for advertiser and is it worth for them, to invest in additional resources to handle big data? Where does all that data come from? Who is storing and analyzing the data and is it compliant with existing data privacy? How far will automation lead us? Will automation replace humans someday or will it just make our lives easier and more successful to deliver better results, as they always say? In fact automation of media buying and the usage of consumer data will continue as a process which has already begun for quite some time.

K. Brosche (✉)
TheAdex GmbH, Hanauer Landstraße 187, 60314 Frankfurt, Germany
e-mail: kolja@theadex.com

A. Kumar
Cadreon, 100 West 33rd St., 9th Floor, New York, NY 10001, USA
e-mail: arun.kumar@cadreon.com

1   Understanding of Data Driven Online Marketing

Data driven online marketing means buying ad space with the help of existing data sources (realtime data) and technologies, to better target the advertiser's audience and be more relevant for consumer.

The objective is to invest the budget as efficient as possible, by finding the right people, who fit the advertisers target audience best, who are already existing customer or who are particularly affine for the applied product or service. At the same time it is about avoiding contact with users, where the advertisement is unwanted or inefficient. As a result, the advertiser will also save costs. That may affect people which are not using the product in general, are using competitor products or, in the best case, already bought the product recently. If that is the case, then it can be of higher interest for the advertiser to re-address this person with advertisement, especially for cross- and upselling purposes.

All marketing is expected to be driven by data; at least that's the promise. In reality though for years, marketing has struggled to integrate various sources of information and provide the right messages to consumers at opportune times. And the feedback loop powered by analytics has often been weak and only allowed advertisers to evaluate the results of their efforts at a broad high level, without providing any granular insights into what actually worked.

In the first few years after the Internet started to grow, data was still scarce and rarely used at scale. Google built the first commercially successful data driven model around the search business. It ticked a lot of boxes necessary to create a data driven ecosystem:

1. Standardised data sets.
2. A marketplace defined by standard metrics.
3. An easy to use interface that gave users access to data in a form that was easily useable.
4. Educating the market on a wide variety of use cases from the simple to the complex.

The search model allowed marketing to bring together a robust intelligence gathering system with an application in as real-time as you wanted it to be. Given the influence of broadcast media over the years and the influence of traditional advertising research with its theories of reach and frequency, it was a big achievement to move the goalposts.

This model heralded the arrival of more sophisticated data driven models and in a classic case of the tail wagging the dog, today marketing dollars spent online have the capacity to help inform and get better returns from offline marketing channel investments and in many cases have supplanted the broadcast world.

Before we go ahead, let's pause and reflect on the word data. The word itself comes from the Latin word 'datum' and means "A thing given or granted; something known or assumed as fact, and made the basis of reasoning or calculation, an assumption or premise from which inferences are drawn." (OED, Vol 4, 264).

The real value only comes towards the end of the definition... "from which inferences are drawn". The quality of those inferences in a way determines the value of the entire data set.

And hence arises this fundamental supposition: Data by itself has no meaning or relevance. The best analogy would be to equate money to data. Money by itself has no value or relevance; put it in a market with context and it's power increases exponentially. What a currency can buy you decides how much you value it or don't. Similarly data is only valuable if you're able to convert it into genuine intelligence that is scalable and useable.

In today's digital world where billions of impressions is no longer just a figure of speech, the true import of the supposition becomes clear. Having data by itself or even having organized it and stored it in a fashion where it is easily catalogued and tagged does not have importance. Data collation and organization should not be mistaken for data driven marketing.

When done right, when data is truly transformed to actionable intelligence, marketing achieves its promise. The benefits are almost immediate.

1. Ability to understand audience definitions beyond syndicated research panel definitions.
2. Ability to follow a consumer across a purchase path that is no longer linear and static but one that resembles a child's squiggles.
3. Ability to not serve messages to consumers who are either not in the right stage of the purchase process or are not as valuable as others.
4. Ability to dynamically alter what one is willing to pay for a consumer depending on a number of variables.

All in, true intelligence can calibrate marketing spend to focus only on the most valuable consumers at relevant times when they are receptive to brand communication. And invariably the feedback loop is equally powerful allowing a marketer to understand whether marketing is hard working or not.

1.1 Where Data Driven Online Marketing Initially Started and What the Benefits Are

So where does the initial idea come from, to refine media buys by using realtime data that originates from CRM data, user interactions or behavioral data? It might have been an idea during a business conversation between an advertiser and a technology service provider, where the advertiser raised the question, if it wasn't possible to get to know their customer much better. What if they knew, which of their customers already bought one of their products, which product this customer is interested right now or in the future and especially knowing exactly when. Then the advertiser just needs to find this customer again to show him the right ad at the right time. In general most of this data already exists somewhere, the challenge is to connect this information with the possibility to buy ad space.

This is not just a pure digital idea. The origin was formed, as mostly in the digital marketing, years before in classical media. Huge data sources, set up by the retail industry decades ago, collected historical purchase data from their customers for years and analyzed it to create and build different audience segments. The newly created data sets have been used for their marketing activities but were monetized as well by selling these data sets or parts of it to other advertiser in the market. A new business model was created and as a result, companies emerged that specialized in collecting, storing and analyzing this data for marketing purposes.

The challenge now is, how can this be done again in the digital world and isn't it possible to even improve this significantly by using realtime data? How can we collect online interactions and e-commerce transactions in realtime, to increase efficiency in customer communication and generate a more attractive ROAS (Return on Advertising Spend)? There are many existing data sources, besides purchase and purchase intent data, which, after being combined with customer and user information, hold enormous potential and allow a more creative and efficient user communication.

2 How Does Data Becomes Enriched for Online Marketing (Definitions)

There is a couple of different kind of data definitions to generate a more efficient ad buying and to make the right buying decisions, by refining media inventory. To fully understand which kind of data is relevant and meaningful to use in different situations, it is necessary to describe these more detailed. In particular there exists three different kinds of data: 1st-, 2nd- and 3rd-party data.

2.1 Differentiation Between 1st-, 2nd- and 3rd-Party Data

"1st party data is information collected directly and stored by website publishers, retailers and other types of companies about their site visitors or customers." Because companies with this information have a prior relationship with their customers, they are able to use this first-party data – which may include names, addresses, phone numbers, site-interaction data and information about products purchased – to communicate directly with them." According to AdvertisingAge "First-party data is what is stored in customer-relationship-management and loyalty-program databases."

Search requests and search behavior of online users in Search engines, and especially the subsequent activities or transactions, deliver another very important data source. If a user clicks on a sponsored link in a search engine and is linked to the advertisers website, the following activities are highly important for successional online marketing activities. It is worth mentioning, that all activities after the click on the sponsored link, means we are talking about 1st-party data, whereas the search information is 2nd-party data. Does the user look at the promoted product or other products on the advertisers website, does he download a pdf file with product information on his computer, does he watch a product video or puts a product in the shopping basket which he then purchase online or not for whatever reason. All this is highly important information for advertisers, which affect all following, data driven online marketing activities.

You can watch an increasing interest of advertiser in user behavior and user interactions with companies, brands or products on social platforms, during the last few years.

The gathered customer insights, based on user behavior on these platforms, regardless if we look at an advertiser webpage on Facebook or a company presence on an business platform like LinkedIn, is highly relevant for data based online marketing.

Data held by publishers collected from their sits or assets is highly valuable and relevant, but not very easily available directly to a marketer. Most publishers prefer to leverage data for their own benefits and do not see the value in selling it in an open market. A far better proposition for them is to work with selected advertisers and match their data sets with the advertisers 1st party data. In some markets, 2nd party data has improved effectiveness in the absence of a 3rd party data marketplace.

All other data, relevant for data driven media buying, which does not contain the already mentioned possibilities, belong to the last option and is called 3rd-party data. These include socio demographic data like age-, gender-, location-data, psychographic data like behavior-, opinion- or interest-data.

There is a great demand for location-based data when it comes to promote a local product (or brand) or a service, which is just available locally. All consumer transactions or purchase intent data are eligible information for data driven ad buying, comparable to the importance of the buying history, collected from retailers, years ago. The information about the last products their customer recently bought is obviously most important for advertisers coming from the e-commerce industry, for driving cross- and upselling. To avoid that a competitor offers a better product to the advertisers customer, it is crucial for them to know when the last purchase was. If the advertiser is aware of that specific information, he can contact the customer again at the right time with the right offer. To increase customer satisfaction and loyalty, the advertiser can create a special offer for existing, loyal customer. This information is not just relevant for e-commerce clients, but for all advertiser offering their products and services online.

2.2 Several Ways of Data-Collection

There are many ways to collect smart data for data driven online marketing. The following Table 1 shows the different kinds of data and gives an overview of 1st-, 2nd- and 3rd-party data, including examples for each one.

Table 1 Differentiation between 1st-, 2nd- and 3rd-party data including examples

1st-party data	User data collected from publisher or advertiser, CRM/newsletter, onsite data gathered from publisher or advertisers website, web-analytics data, client database (on- and offline)
2nd-party data	Collected from advertising campaigns from various channels: i.e. paid search, display/bannering, mobile, paid social, social advertising, video, TV spend and media information, search behavior in search engines, behavior and interaction on social platforms
3rd-party data	Data from 3rd-party sources i.e. e-commerce data from various platforms: purchase information, purchase intent data, product interest. Other 3rd-party data sources like weather data, geo-location, socio- and psycho-graphic data, online surf-behavior (publisher data)

2.3 Data Collection Following the Rules of Data Privacy

Let's look at a very important topic, when it comes to data collection, storing, analyzing and processing of user data: Compliance with existing privacy protection law. Each company has the obligation to follow the local privacy regulation. Today there is no global privacy regulation. Each country has its own understanding and its own law. In the EU this law exists in a revised version and gets constantly reworked in accordance to the changing requirements for companies as well as consumers.

Internationally, the application of the privacy laws has not been consistent. In the United States in particular, a fair amount of freedom is given to publishers and advertisers to collect and leverage data for targeting. The definition of privacy is given a fair amount of latitude. For example, while Facebook is allowed to leverage user data for targeting, it does face a fair amount of scrutiny from the US Government when it tries to leverage that outside the walls of FB. This gets even more complicated in the mobile environment where telcos and publishers are privy to information such as location. A fair amount of self-regulation helps the industry where players like Google and Facebook establish a few ground rules in consultation with government agencies.

In APAC, the laws are quite fluid and do not offer the same level of protection to consumers as in the Western World.

3 Which Advertiser Will Benefit From Data Driven Online Marketing

It is comprehensible that advertisers are questioning the efficiency of the use of external data to enrich their media buys, by adding additional costs. But there is more to it than just adding costs for the use of 3rd-party data. The need to facilitate 1st-party data in an external technical environment like a Data-Management-

Platform (DMP), also additional investments providing resources to analyze and segment audience profiles as well as a consolidation with CRM data and any other 2nd-party data, all this generates costs that should be considered. The fast technical development and the possibility to access media inventory (ad space) in realtime, will provide advertisers with phenomenal opportunities and leads to a point where advertisers will run their online marketing activities data driven. Most of the advertiser collect CRM data in-house with an own tool or get it as a service from external companies. Nearly all of the generated campaign activities and interactions from online marketing campaigns are stored in any ad server. The activities, interactions and purchase data from paid search activities are stored in another data source. This may be stored in-house or like most of the clients, this information is stored with the technology provider directly. The challenge for advertisers now is to consolidate these data sources, structure the existing data and make them useable for the media buying in realtime – across multiple channels.

In general almost every company owns data from one or different sources (1st-, 2nd- and 3rd-party data) and it is highly recommendable to use this data for marketing purposes, this of course under strict compliance with the effective data privacy regulations within your market.

3.1 The Value of User Based Data for Advertiser?

Very often the question comes up, how much it is worth for the advertiser, using profile-based data for online campaigns. But there's no 'one fits all' answer. Depending on the advertiser and his products, the volume of money he spends for online media, the generally use of targeting of the advertiser, the goal the advertiser wants to achieve with his campaign, all this questions are important when it comes to define, if additional costs for user data are worth it. It is highly recommended for an advertiser to firstly begin to track data of all three kinds (1st-, 2nd- and 3rd-party data), consolidated within one platform, which is, secondly, able to make this data available for addressable media buying in realtime. If not all generated data (e.g. advertising contacts with user or customer) gets collected and stored in one platform, the advertiser just look at a snapshot and possibly spend money to target users, who already had contact with the campaign or bought the product via another channel (e.g. mobile). This leads to wrong assumptions where the traffic or transactions came from and wrong decisions where to spend more money in further campaigns. Without a holistic view through an overall tracking, the effectiveness of each channel and the whole campaign is not correct. How high the value is for a single advertiser can just be calculated in-house, but in fact all campaigns will be data driven and monitored with a holistic view in the future.

3.2 Is It More Efficient, Taking Higher Costs by Using 3rd-Party Data for Programmatic Advertising, to Leverage Campaign Performance?

Watching the consumer journey in more detail and with a multi-channel view is highly recommended for each and every advertiser with no exception. There are a bunch of success stories and case studies from each industry. Because of great differences between the product-groups themself and even bigger differences between the specific industries, the results of the cases can't be compared with one another. But what can be said overall is, a consistent tracking of user interactions across all touch-points, combined with programmatic buying, leads to a significant efficiency uplift, especially for high involvement products. Not just the use of 1st- and 2nd-party data is very important for the success, but external data, 3rd-party data sources, can be of major interest.

> It is of major interest for a travel agency, a tourism-board or an airline, if they can target an audience, which already booked a flight in the last 12 months to foreign countries or rent a car. At the same time these advertiser want to exclude people, which already booked their holidays, a hotel, a flight or something similar in the last 3 months.

4 Requirements for Analysis and Processing of Realtime-Data (From a Technology and Infrastructural Standpoint)

To be able to handle multiple data sources, consolidate, analyze, restructure and process these data streams, there are a lot of to do's an advertiser has to take care of initially, before using these data sources for programmatic buying in realtime. Besides allocating experienced, qualified staff, there is a need for a technical infrastructure to provide a Data Management Platform (DMP) beforehand. There should be an internal agreement in place first, whether to setup the technical infrastructure in-house or to work with an existing technology provider. The partner of choice should have the ability to deliver solutions and services to fit the company's requirements in the best way.

Some of these requirements might be:

- Delivering a stand-alone solution / independent from inventory platforms but interoperable with relevant platforms
- Capable for multichannel tracking
- Interoperable and / or already connected to 3rd-party Data-Exchanges
- Data-privacy compliant with specific market / country
- Technical accessibility ("Uptime") and support level
- Willingness and ability to customize the technology solution (to realize certain client wishes) in terms of data-matching, audience segmenting, create statistical twins, algorithms, and so forth

These are just a few examples to look at, when it comes to find the right DMP or technology partner. Flexibility in terms of existing (API's) or to provide a connection to internal or external data sources, software or systems are very important criteria to find the right partner, too. The technology partner should allocate resources for that important topic. It might make sense to work with one of the existing, qualified consultants in the industry to support you, to find the right partner.

5   Outlook: Data Driven Online Marketing in the Near Future

Depending on the specific market and how far the market is developed with regards to inventory- and data-accessibility, the media volume for programmatic will increase rapidly or moderate. There is already a shift of advertising spend to data driven marketing. It just depends on the market readiness. The more data is available for data driven media buying, the more ad space will be accessible for advertiser and the more money will be spend in this area. Advertisers are actually working on consolidating their data in a single platform to be able to analyze and process their data multi-channel to get a holistic view of the customer journey and to place ads in front of the right audience at the right time. There are just a few advertisers who combine data driven online marketing with the use of dynamic creative ads yet. There is huge potential to increase the campaign efficiency, if the messaging will be done automatically (dynamically) as well.

> Let's say an advertisers goal is to drive interactions (e.g. click on an ad) or drive conversions (buy a product online). The algorithms learns from activities and interactions of earlier campaigns and decides in realtime, depending on the advertisers campaign goal, which user sees the ad, at what time and in which environment the ad will be placed. This could mean that a user who already had contact with an ad, and is part of the target audience, sees the ad more often than others, until he interacts with the ad or buys the product. The algorithm controls the messaging and can modify the communication, based on the customer's behavior and his interactions with the ad. Did the customer already buy the product, he won't see an ad for this product until the product lifecycle starts again. Because the advertiser knows when the customer bought the product, he is able to contact him at the right time again, maybe with a special offer for regular customers.

The part of media being bought directly and manually without profile based user information will decrease in the coming years and will be purchased more and more automatically. Publishers make their inventory available in Sell Side Platforms (SSP's). Agencies and Traders can access to this inventory via Demand Side Platforms (DSP's), which are connected to the Sell Side. This process is not just more efficient but also enables data driven marketing in realtime.

5.1 The Role of Media-Agencies and Media-Planner in the Future

The fragmentation of data has disrupted the entire advertising ecosystem. In the traditional model, the buy side represented by agencies and marketers held the upper hand. Advertising research, recall studies, panel based or syndicated databases provided knowledge to agencies that helped them understand the value of specific media channels or audiences in granular detail. The talent bases were at worst balanced and at best biased towards the buy side.

The disruption wrought by the programmatic ecosystem has meant that the power of large data sets with the ability to execute now rests with the buy side. And as more channels, both digital and traditional begin to rely on server based mechanisms, that shift will only get more pronounced.

Compounding this problem is the opacity that exists in programmatic buying today. The lack of trust in the digital ecosystem has meant that in some cases the buy side is fragmented; agencies and their clients are not always on the same side and many ad tech players have rushed to fill this void and build direct relationships with advertisers.

The noise generated by issues like viewability and fraud has obscured the very genuine advancements that have arisen in the ad industry where media has become addressable. The concept of data powered marketing or buying was limited to the programmatic space even till a year ago but today there is real potential to change the ad model completely. Evidence of this lies in the embrace of data management platforms by broadcast networks in the US who have slowly woken up to the potential of their audiences and understood the value of that understanding.

The agency model that has forever been in flux has experienced more of it in the last 2 years than in the previous 10 years. And with that the role of the media planner has come under scrutiny. The blunt reality of today's world is that buyers have greater intelligence at their fingertips than do planners. While planners still rely on panel based surveys and sampling, buyers work with larger data sets and get access to them in real time.

This has almost bought the agency full circle. Go back 20 years and before the rise of the AOR or the agency of record, planning and buying were not separate functions. As complexity increased the functions dispersed. However with ad tech, it is possible for an individual to get a real time assessment of the marketplace, estimate the value of an audience for the business and leverage any data he or she might have to make decisions. Should one compromise agility for slightly more depth?

One solution in the interim to bridge the gap is to democratize the data and intelligence that rests in a buyer's realm to planners. Data driven planning can leverage the rise of DMPs that can give access to information in a structured format across the agency and the marketing organization.

In the long term, it is tough to see split functions. Planning will morph into Strategy and the role of the media planner will become obsolete. Trading will also witness a change, as strategic buyers who create marketplaces will be the norm whilst the traditional planning and buying functions will merge. This will be the ultimate gift of data driven marketing leveraging the power of automation: smarter and fewer human beings operating in a more agile fashion.

6 Conclusion

There's no doubt that the amount of existing data and the volume of processed data through professional technology companies and advertisers directly, will increase massively. Through the availability of this data, the generated insights and the possibility to process this data in realtime to optimize consumer communication, data driven marketing will raise significantly, now and in the next years. The current possibilities to track and optimize multi-channel activities, is another accelerator for data driven marketing and offers a more precise picture of the customer journey. Offline data providers transfer their database to online by utilizing DMP's and enrich existing online profiles to gain a more granular picture of existing customer profiles.

In the next years we'll see the merge of "digital screens" with "offline screens". They will then all be "connected screens", regardless if we carry them in our pockets or we see them outside in the streets in the form of city-lights posters or interactive screens on railway platforms (digital/connected out-of-home screens). They will communicate with our mobile devices, which trigger the spot or message we see at the screens, depending on our (anonymized) user-profile. Our mobile devices serve as a GPS recipient to deliver our geo coordinates and actual personal interests, e.g. interests in products or services, or buying intent. Algorithms decide which message we'll see, based on our actual life situation, our health information (relaxed or stressed) and our user behavior. The current user location decides about the screen on which the user will be contacted and where the messages will be sent.

Further developments will include the usage of mobile devices and wearables to collect more personal information. The connection with the internet will add more and more data sources and therefore more possibilities how to enrich certain user profiles. TV and video advertising is about to merge right now (See also Scott Ferber 2014). The industry just talks about screen planning and the simultaneous usage of multiple screens.

The usage of anonymized, profile-based consumer data for media buys today is a fast growing business. It will benefit from the merge of the different channels through fast developing technologies and the consolidation of data streams within a few platforms. There's no border between on- and offline, everything is online and connected. The Internet of things will connect all devices and things of our daily life and permanently transmit data to companies. The customers will benefit from a much more relevant and beneficial communication between them and their preferred brands and products.

Bibliography

AdvertisingAge, Published on 03 December 2013 (Online) http://adage.com/article/glossary-data-defined/party-data-defined/245054/
Bundesverband Digitale Wirtschaft (BVDW) e.V.: http://www.bvdw.org/english.html
BVDW (Hrsg.). Realtime Advertising Kompass, Publication date 2013/2014, http://www.bvdw.org/mybvdw/media/view/realtime-advertising-kompass-2013-2014?media=5071
Ferber, S. (2014, February). How will marketers buy and sell media in 2020? (Online). http://www.exchangewire.com/blog/2014/02/11/how-will-marketers-buy-and-sell-media-in-2020/. Accessed 30 Mar 2014.
O'Leary, N. (2014). Matt Seiler is out to remake and automate the media agency world (Online). http://www.adweek.com/news/advertising-branding/matt-seiler-out-remake-and-automate-media-agency-world-155047. Accessed 26 Jan 2014.

Kolja Brosche is working in the online marketing industry since over 16 years, working in different roles on agencyside and the salesside before he recently moved to an Adtech company. After building up Cadreon (IPG's Trading Desk) and Mediabrands Audience Platform (MAP), both member of the Interpublic Group of Companies, in Germany, he moved as a COO to The Adex, a technology leading Data Management Platform (DMP). A DMP enables customer to collect, aggregate and analyze data generated by user behavior in the internet and make it available for customers to create data segments with their own data (1st-party data) and combine it with available data (3rd-party data). Customers can use these created segments and buy media against these data sets by using a Demand Side Platform.

Furthermore Kolja is being involved in the focusgroup "Programmatic Advertising" within the BVDW, as well as speaker on panels and conferences such as the ATS or the dmexco.

Arun Kumar is IPG Mediabrand's Global President for Cadreon, the group's programmatic buying devision. Before he headed Mediabrands digital business in APAC and lead the Mediabrands Audience Platform across the 14 most profitable markets outside the U.S., with diversified digital disciplines including Search, Social, Mobile & Exchange/DSP based digital buying. On the way up to IPG's top he hold Management Positions with Starcom MediaVest, Isobar, Aegis Media and PHD in APAC.

Redefining Retargeting

Grégory Gazagne and Alexander Gösswein

We all know the volume of data is exploding and the need for marketers, publishers, and their stakeholders to manage it, has become a bigger issue than ever before. Imagine: Today, the digital data available worldwide is 4.4 zettabytes. Just to visualize: If you wanted to store this amount of data on 32Gb iPad Airs, you'd need a pile as high as two thirds of the distance to the moon. And you could save music in 4.4 zettabytes with a run-time of more than eight billion years. According to estimates by market research firm IDC and data storage manufacturer EMC, in just 7 years this is expected to grow to 40 zettabytes (IDC/EMC Corporation: The Digital Universe in 2020. http://www.emc.com/collateral/analyst-reports/idc-the-digital-universe-in-2020.pdf (2012)).

With this evolution has come an enormous opportunity for display and those who know how to make data actionable. Advertisers have an increasing amount of data at their disposal that can provide real added value when properly evaluated. Same with publishers: By integrating data in a reasonable and cost-effective way, they have the opportunity to support advertisers' campaigns; providing inventory with a higher value means: they can get better prices for that inventory.

1 It Is All About Big Data, Big Insights: And Big Efforts

Data is often the barrier for many companies to enter or further establish their involvement in the performance advertising market. The main consideration is the huge effort needed. Alongside the capture and storing of data, you also have to

G. Gazagne (✉)
Criteo, 32 Rue Blanche, 75009 Paris, France
e-mail: g.gazagne@criteo.com

A. Gösswein
Criteo GmbH, Gewuerzmuehlstr. 11, 80538 Munich, Germany
e-mail: a.goesswein@criteo.com

visualize and analyze it. Because the important question is not about Big Data – it is about Big Insights and how to get them.

It is not just about additional technology and staffing resources. It is also a matter of time – and lots of it. Any company able to leverage huge amounts of data has invested years of time and people, dedicated to just this one goal: Making data usable. It is a consistent period of testing and learning. And sometimes failing and testing again. That's where specialized suppliers add value: They are able to remove the complexity and thus ease the way for advertisers and publishers; so they can focus on their core business.

2 Success Factors for Programmatic Performance Marketing

The critical factor for successful programmatic advertising is a significant amount of data that is both scalable and relevant. But, it is also about sophisticated algorithms, a customizable ad design, and a tracking system that makes it possible to draw the right conclusions and make iterative adjustments to the campaign set-up. Showing the right ad to the right user at the right time requires the intelligent use of hundreds of variables. And – this is critical – the algorithms need to automatically and continuously learn from the data processed – otherwise the users most likely won't see the most interesting and relevant product at a certain time.

2.1 Intelligent Algorithms

When talking about performance marketing, we are sometimes asked why we don't just call it retargeting. The reason is basic enough: simple retargeting, in which users are only served ads for products that they previously viewed yet did not purchase, is not only outdated, but is also not attractive – nor expedient – for advertisers and consumers. In fact, the banners used in this method do not even need to be created in real time.

To meet both the increased demands of users and the goals of advertisers, i.e. to ensure higher conversions and a good return on investment, much more is required – namely real performance marketing. And the engine today is a critical part of this equation.

Good engines contain globally consistent machine learning algorithms to generate predictions and recommendations, that makes it possible to operate in a scalable manner on a global scale – and in real time. Good algorithms not only learn from historical data, but also continuously incorporate new information. This improves the accuracy of predictions and recommendations with every ad delivered.

Systems need four components: •prediction, •recommendation, •bidding, and •dynamic creative optimization. Prediction algorithms select the right audience, the right advertiser, the right medium, and the right ad space at a particular moment, while recommendation algorithms determine the most relevant products to display.

Bidding algorithms calculate and optimize the value for each user in real time in order to determine the maximum bid price. And lastly, dynamic creative organization creates the right look and feel for the banner.

All of this information can be fed into demand side platforms and used to manage ad campaigns.

2.2 Customizable Ad Design

The creatives must also be produced in real time. Within six milliseconds the banner – that most appeals to the user – needs to be created from millions of banner variations. The objective here is not to push the boundaries of creativity, but rather to be as appealing as possible in order to capture users' interest, thereby enticing them to click on the ad and returning them to the advertiser's product purchase page.

You often hear in industry discussions that advertising in general – and banner ads in particular – need to get better. By this, almost always "more creative" is meant. However, definitions of "good" and "bad" banners differ, depending on whether they are viewed from the standpoint of performance or of creativity. What the designer thinks is attractive often does not work as well from a performance perspective as supposedly less creative banners. The challenge is to create the optimal banner for the advertisers: well performing yet consistent with their corporate identity and highly appealing to consumers.

The real question around creatives is: Are they customizable to any specific situation? Imagine you have a set of great creative banners: Do they adapt to any screen the user you target uses? Do they work when you expand your business to new markets, where completely different size formats for your ads are required? And can you easily switch from banner to text ads if the environment requires you to or the performance is more promising?

Customization is key in Performance Marketing today. It is all about extending the number of possibilities of the creative, to fit different screens and usage behaviors. On top, it is important to optimize on the kind of ads delivered to an individual user; those ads which are most appealing to each user, are of course the strongest in performance.

2.3 Tracking Success

Along with the specific impact in relation to the pursued goal, a campaign's success is measured by its performance – in other words, to what extent it achieves broad, scalable, and consistent reach and the desired sales volume. It is important to continuously monitor performance indicators in order to identify peaks and dips and make any necessary adjustments to the campaign. The corresponding data can be translated into actual insights and incorporated into future optimization measures.

Rigorous tracking and analytics make it possible to show metrics such as overall performance, channel-specific performance, and, if applicable, partner performance over time, while also identifying optimization potential with respect to conversion and resource allocation per channel. Reporting dashboards, which aggregate the real-time analysis of all relevant data and are customizable to advertisers' needs, enable the real-time monitoring of campaign results. This allows advertisers to see at a glance how each cost-per-click (CPC) bid is influencing individual categories and overall campaign performance, while the engine optimizes against the predicted likelihood of sales in the back. As a result, advertisers have a constant overview of the campaign and can make appropriate changes where necessary.

3   The Importance of Precisely Defined Goals

Advertisers' goals can be as diverse as the advertisers themselves. They can range from generating maximum revenue, through promoting the sales of particular product categories, to targeting new segments of buyers.

But the core issue is always the same: What is important to the advertiser? It is here that it is necessary to help advertisers best achieve their goals and define corresponding performance indicators. This is especially crucial in difficult economic times when budget cuts are a realistic concern. Marketing activities capable of delivering sales growth are not only immune to cutbacks but are also of invaluable benefit well beyond the marketing department.

That aside, performance display advertising is exceptionally well suited to meet the current needs of advertisers. The chief concern here, as previously mentioned, is how to most effectively and most securely use the data that already exist in-house. Doing this is usually too technically demanding and too far removed from advertisers' core operations for them to take on the task themselves.

Let's look at an example: An advertiser is particularly interested in targeting a certain segment of buyers – perhaps frequent, big-ticket, or occasional shoppers. All of this information is usually stored in the CRM system. If the advertiser categorizes this data properly (e.g. Category 1 = Big-ticket shoppers), it can send the information anonymously to the service provider and manage the Programmatic Advertising campaign accordingly.

But does that mean advertisers shouldn't invest at all in analytical people? No, it is important, even vital for them to staff up quickly with people who understand data. Who are able to analyze and visualize it and thus effectively help integrate it into their campaigns. These people 'think tech' and interact this way with your providers to increase your Return on investment (ROI) to the max.

4   The Art of Finding the Right Attribution Model

Measurability and accountability have been every advertiser's Holy Grail since the late nineteenth century, when US department store merchant John Wannamaker famously quipped: "Half of the money I spend on advertising is wasted; the problem is I don't know which half." For the past few years, marketing attribution tools and techniques have brought us closer to fulfilling the quest – but we haven't quite reached that goal just yet.

As the Criteo research paper "Marketing Attribution Comes of Age"[1] shows: The majority of advertisers still use last click as their primary attribution model – even when they consider it insufficient. There are still many organizational constraints and technical challenges that stand in the way of widespread adoption of multi-touchpoint models. But we also learned some valuable lessons with the paper which are:

Lesson #1: Challenge Whatever Your Model Tells You A model always remains just that: a simplified version of reality used for practical purposes. Simple shouldn't mean simplistic, though. In particular, make sure that you get a clear view on what touchpoints your model is missing. If you don't, you might miss sales.

Lesson #2: Stop Assuming, Start Testing The best way to demonstrate causality is by testing. If you have doubts concerning the value of a marketing channel or if you would like to see what doubling the amount you spend on it would really do to your sales, the best way to find out is to set up a test.

Lesson #3: Focus on Touchpoints That Truly Have Influence on the Purchasing Ecision Many touchpoints serve only to help a user navigate from one place to another; they don't actually influence any buying decision. Removing such "navigational" touchpoints from your model is less game changing than switching to a new attribution model, but it has a big impact on your results – and on your ability to generate more sales.

The ability to track users across multiple devices is perceived by this research as one of the most important challenges faced by the industry today. The explosion in Mobile usage is one of the main reasons for this.

5 Mobile Is the New Black

The time users spend with mobile devices is increasing heavily. Today we spend already twice as much time with our smartphones and tablets than just 2 years ago. This is due to the spread of mobile devices. Of the world's 7 billion people, 59 % will use smartphones in 2015, according to eMarketer.[2] Tablet use is also on the rise, with 1 billion consumers worldwide using tablets in 2015, up from 660 million in 2013.[3]

---

[1] Criteo: Marketing Attribution Comes of Age. http://www.criteo.com/media/1024/research.pdf (2014).
[2] eMarketer: Smartphone and Tablet Penetration in Select Countries, 2015–2019 (2015).
[3] eMarketer: Tablet Users Worldwide, 2013–2019 (2015).

But people don't buy on mobile, right? Well, that's the same estimation as in the nineties, when people were sure that no one would buy on the internet. Today, we know the true story. With mobile it is happening even faster, as the Criteo "State of Mobile Commerce Report"[4] shows: mobile already accounts for 34 % of eCommerce transactions globally.

The need to have a web commerce store optimized for mobile users is thus bigger than ever, in fact it is a critical requirement. But also companies whose users are more used to apps have new opportunities now. Until recently, a lack of technology hampered marketers' efforts on mobile. But three recently-developed technological advances are helping them to monetize these valuable user bases:

Advance #1: In-App Advertising Being able to immediately display a dynamic, personalized ad within an app enables marketers to offer the kind of highly relevant consumer experience that is so crucial to conversions. Advertising companies have invested significant time and money in developing mobile apps and user bases. Thanks to in-app performance advertising, they can now deliver the same highly personalized and relevant ads to in-app consumers based on their browsing habits in the same way that they can with desktop users.

Advance #2: Dynamic Ads There are two key issues that plague the majority of today's mobile advertisements. First, they are largely impersonal and frequently irrelevant. As a result many users often completely ignore in-app ads entirely. Second, mobile ads can be difficult to interact with as they often feature a vague call to action, along with branding, and just link to a company's homepage. This increases the likelihood that the user will stop engaging with the app and the initial click will be wasted. Whereas, highly targeted dynamic ads customized to the users' individual experience, provide a significant increase in value and boost the probability of a conversion.

Advance #3: Deep-Linking Mobile deep linking creates a direct link to any location within an application. For example, if a consumer was previously viewing a tennis racket, a dynamic ad will be displayed showing a specific offer and after clicking will lead him directly to the purchase page in the sports company's app. What makes this so important? This process substantially reduces the number of network hops and redirects, providing the user with immediate access to specific content.

And so we see that engaging the mobile user is possible today. The challenge has become managing and balancing a desktop and mobile strategy as the usage of desktops remains stable and mobile continues to grow. So marketers don't have to master a shift from one channel to another, but are facing a complex scenario.

---

[4] Criteo: State of Mobile Commerce Q1 2015: http://www.criteo.com/de/resources/mobile-commerce-q1-2015/ (2015).

6 User Engagement on Multiple Devices

Today's shopper often uses a laptop, a tablet and a smartphone. By 2017, each user is expected to be connected through 5 devices already.[5] Complexity is constantly increasing, in terms of numbers of devices used as well as using behavior. The majority of people – e.g. 90 % in Germany[6] – use tablets mostly at home and thus at different times than smartphones and PCs. It is getting more and more unusual for a shopper's decision-making process to involve just one device and one online touch point. Their brand interactions are changing as well. Consumers want a consistent experience when they move from device to device. In fact, some say they will stop interacting with a brand if the experience varies too much across devices. Overall, consumers view a brand's websites, apps, the retail store and ads as part of the same experience. This highlights the growing need for marketers to have an effective cross-device strategy to be able to meet customer expectations and to optimize their ROI on advertising.

Technically, for advertisers an exact match solution – one that will help them figure out if the person who saw their ad on a tablet is the same one who made a purchase on their PC later that day – will be critical to cross-device success.

But why is exact match so important? Unlike implied match, which is a machine-learning model that uses device characteristics to estimate a probability that two users are the same person, exact match uses unique identifiers to precisely match consumers, resulting in an accurate match. Implied match can create false positives by matching two different people and its precision varies, ranging from 40 % to 75 % – resulting in a much lower performance/ROI. With exact match solutions, digital advertising becomes more efficient for the entire ecosystem.

7 Data Protection and Privacy: Personalized Ads as a Service for the User

Of course, some consumers are unsure of targeted advertising. Their main concern is: "How do they know?" Well, Performance Advertising in general and Cross-Device Advertising in particular are smart technologies, but no mystical mystery. They are a service for consumers, which should be combined with highest privacy standards, transparency and user choice.

7.1 Why Ad Blockers Are Not Useful for the User

Since the dawn of advertising, there has been some resistance to advertising, particularly in regards to advertising on the 'cost-free' internet. Users are very

---

[5] Cisco: Cisco Visual Networking Index (VNI) Global IP Traffic Forecast (2013).
[6] Tomorrow Focus: Mobile Effects 2014-I: http://www.tomorrow-focus-media.de/uploads/tx_mjstudien/TFM_MobileEffects_Studie_2014-I_01.pdf (2014).

sensitive to anything they feel is intrusive and irrelevant to their experience and this is where the popularity of ad blockers has grown from. But there is a major problem here: The internet is not free. People who create quality content usually want to get paid for it. Thus advertising is needed. Without companies paying publishers and content providers to show their ads, users would have to pay to access webpages they are interested in. That has always been the rule of the internet. And it works.

The arrival of ad blockers has essentially broken this rule. Users think it is in their interest to be protected from ads. But what they don't see is that ad blockers kill the profitability of content providers, which in the end will lead to the death of the 'free' internet: If Advertising companies don't subsidize the content; then who does? Inevitably it will be the users through some sort of payment for access and this is not a sustainable model.

As soon as users understand the implications of ad blockers and delete them, the question arises as which ads they want to see: Mostly irrelevant ones? Or ads that are highly relevant and useful to them? Ads that provide them with product recommendations that fit their interests at that time? Most users will opt for relevant advertising. Having well-performing personalized ads – ensures ad providers can pay publishers better for their inventory than advertisers with standard ads for everyone. This results in return again in increased and better content they can provide. In the end: Isn't this what we all as internet users want?

7.2 Transparency and User Choice: Key to Personalized Services

Data protection was an issue of intense debate even before the NSA scandal. The protection of consumer data is regulated differently around the world – sometimes more, sometimes less stringently.

Before examining the pertinent laws in individual countries, it is imperative to first make sure that solutions are user-oriented. This means that users should be well informed at all times and can make conscious decisions about their online experience and about the usage of their data. Users should always have complete transparency with respect to the ads they are served and should be able to specify their preferences via an interactive "i" symbol found in targeted ads. The technology uses information collected from users' previous browsing activity. Criteo for example introduced the "i" symbol in all of its advertisements in 2009 – before the industry began to concretely consider self-regulatory action.

Such an approach has been endorsed by leading industry associations. They believe it benefits the entire industry when there are solutions in place that allow users to make informed decisions about their browsing activity while also allowing advertisers to place personalized advertisements. A suitable method is to proactively inform website visitors about the type of cookies being used and in which way, and to provide them with options to configure these.

8 On Route to a User-Centric Cross-Device Solution

Successful programmatic performance display advertising is dependent on various factors such as the right data, properly functioning algorithms, banner design, and tracking. In addition, advertisers must be able to monitor and manage campaigns at any time. Data protection and data security play a particularly important role in this regard.

Today, sophisticated technologies and algorithms make it possible to display personalized and relevant ads to every user – in real time, across the globe, and regardless of device. But this is not the end of the journey. There is always something new – marketing decision makers should never refuse to adopt new channels. Mobile shows how quickly a trend can become main-stream and what you risk if not going with it. It is not unlikely to imagine a time soon when we will see in the industry a user-centric multi-channel/cross device solution, providing the ability to say: X or Y user reacts on 10 desktop banners, 4 in-app ads, and 3 personlized emails. Based on this information advertisers can better manage frequency and at the same time optimize drastically on Cost of Sale.

9 Conclusion

Advertisers should take multi-device seriously. Track success carefully and trust tech-savvy providers able to handle the exploding amount of data. Additionally it is advised to staff, efficiently and quickly, with analytic controllers and scientists to get the most out of new opportunities.

Grégory Gazagne joined Criteo in 2010 as Sales Vice President and in charge of the international development. His success in opening 12 new markets in only 12 month led him to the position of Managing Director for France, Southern Europe and Latin America in January 2011. This graduate of ISG and former French Navy Commandos started his career at TF1 in 1999. After 2 years spent at IP, he joined Aegis in 2002 as Commercial Development Manager. In 2004 he created the Special Operations and Partnerships Department at Yahoo! France before he gets promoted Head of Sales and Deputy Commercial Director. At Criteo, Grégory Gazagne has been Managing Director Europe since January 2013.

Alexander Gösswein is Managing Director Central Europe at Criteo. Before that, as Managing Director DACH he had been in charge of Criteo's business in Germany, Austria, and Switzerland. Alexander Gösswein has more than 17 years of experience in the marketing and media sector with leading positions at Yahoo! (Overture) and Microsoft Advertising. Before joining Criteo he built up goviral Germany, which has been acquired by AOL Europe in January 2011.

Driving Performance with Programmatic CRM

Florian Heinemann

So far, the systematic scaling of the acquisition of new customers via programmatic advertising has turned out to be associated with considerable difficulties. The main reason is a shortfall in terms of quantity and quality of available data and its limited predictive power regarding actual performance. In addition, accessing and using data in most cases have proven to be rather complex.

Aside from these difficulties, large players such as Google, Facebook, Amazon, Apple, and Alibaba have developed into platforms combining (a) large reach (b) high login frequency, and (c) great depth of data per user across different devices. Due to these factors, these "high/deep reach login platforms" continue to gain importance and lead to an increasingly transparent advertising landscape. Initially, this may appear to be a positive development for advertisers, since all relevant users can be reached via just a few platforms with great targeting precision. Medium to long-term, however, this can impair their position, since it becomes increasingly difficult for advertisers to differentiate themselves from others and maintain a competitive advantage based on the sophistication of their advertising activity. The problems arising for publishers and especially media agencies regarding this development are even more severe. But that will be covered only as a side issue in this article.

Using proprietary customer relationship management (CRM) data of the advertiser in conjunction with programmatic advertising appears to be one approach to mitigate the shift of power away from advertisers to the "high/deep reach login platforms". And there is still a lot of unused potential to be systematically exploited in this area. In addition, the impact of this approach can be expanded and made more scalable by using CRM data to reach potential "upper funnel" customers via "lookalike" techniques.

F. Heinemann (✉)
Project A Ventures GmbH & Co. KG, Julie-Wolfthorn-Str. 1, 10115 Berlin, Germany
e-mail: florian.heinemann@project-a.com

To make systematic use of programmatic CRM in a coordinated approach of addressing customers, the underlying (a) logic/concept and (b) technical infrastructure have to be well designed so that existing customers and potential customers can be addressed in a structured way and across multiple channels/platforms. In order to implement integrated multi-channel/platform marketing, models based on individual user behavior are required to determine – among other factors – (a) what message (b) when (c) how/in which format (d) at which frequency (e) on which platform/channel, and (f) at what price should be placed in order to achieve optimal results. Furthermore, data collection, aggregation, and activation/usage need to be integrated into one comprehensive infrastructure.

1 On the Current State of Programmatic in Performance Advertising

The original assumption that in terms of scalability and systematization, programmatic advertising will work similarly to search engine advertising (SEA) in generating stable performance across most advertisers has proven false. Regarding the acquisition of new customers, systematic scaling in programmatic advertising, at least apart from retargeting, has proven to be complex and difficult while requiring very advertiser-specific efforts. A primary reason is that data available today comes with shortcomings in terms of quantity and quality and often lacks reliability in its predictive power with regards to performance. Furthermore, the cost of accessing data is relatively high and/or unrelated to its predictive power regarding actual performance. In addition, the usage process is often too complex. Advertisers have had to deal with these kinds of difficulties for the past several years.

For these reasons, there is a clear tendency that successful providers of tools in the programmatic space have focused increasingly on brand and premium advertising while neglecting performance.

The exceptions are (a) companies with a clear re-targeting focus such as Criteo and most importantly (b) the large "high/deep reach login platforms", e.g. Google, Facebook, and increasingly Amazon, Alibaba and potentially Apple. Due to their existing user base and high reach combined with frequent logins across multiple devices, they have easy and cost-effective access to large quantities of high-quality data on an individual user level. These platforms are therefore not affected by the problems outlined above. And since user identification solely based on cookies becomes increasingly problematic with the rising importance of multi-device usage, login frequency with such platforms has become a key differentiator with regards to data quality. Due to these factors, the "high/deep reach login platforms" will even increase their already existing dominance of the advertising landscape in the years to come.

At first glance, this increased importance can even be considered positive for advertisers, since the advertising ecosystem is becoming more transparent and all users can be reached cost efficiently using only a few advertising platforms. This advantage is, however, only short-term in nature. Medium to long-term, this trend leads to a deterioration of the advertiser's position with a relative power shift to the "high/deep reach login platforms" that control the access to the users. In advertising in general, but in particular in the context of programmatic advertising, competitive edge is directly correlated to information advantage towards other players in the system, i.e. a state of asymmetric information. In a non-transparent (programmatic) advertising landscape, this asymmetric knowledge distribution regarding purchase prices, proprietary targeting data etc. can be used by respective advertisers to generate a competitive advantage. With continuously fewer large platform players that have a strong interest in a transparent design of their own advertising platform to maximize the auction/bidding revenue on their ad inventory, it is increasingly difficult for advertisers to differentiate themselves.

Publishers and media agencies also face severe difficulties in this regard. The publisher's relative position of power will definitely deteriorate compared to the login platforms, as the context they provide as a targeting criterion is losing importance vis-a-vis data points related to the individual user. Forming alliances to generate and increase their level of "deep reach" on an individual user level seems to be the only option to counteract this shift in power. However, it remains unclear to what extend these measures can compensate/mitigate this development. The structural disadvantage of publishers is enormous. Some publishers have already acknowledged this by handing over the management of significant parts of their advertising space to large login platforms. Another option might be to productize existing capabilities such as storytelling and offer them to advertisers.

Media agencies are certainly in the most difficult position when looking at the current trends in the advertising landscape. So far, they have lived well from providing guidance for their customers in a fragmented and untransparent market and bundling demand. This has allowed them to run a very profitable arbitrage play which will become increasingly difficult with fewer platform players that are highly interested in (a) transparency and (b) directly interacting with advertisers.

2 Generating Competitive Advantage Through Proprietary CRM Data in Conjunction with Programmatic Advertising

2.1 Relevance of Programmatic CRM

A logical consequence of the situation described above is that advertisers have to explore other opportunities to potentially generate a competitive advantage. One option is using proprietary, first-party data on existing customers and to apply this data to programmatic advertising. Through the intelligent use of existing customer data that is in full control of the advertiser, a defendable information advantage can be secured, and thus the increasing shift of power towards the "high/deep reach login platforms" can at least be mitigated.

Looking at the advertising possibilities of the different stages of the purchase funnel, i.e. the various phases of the user/customer journey in terms of (a) the

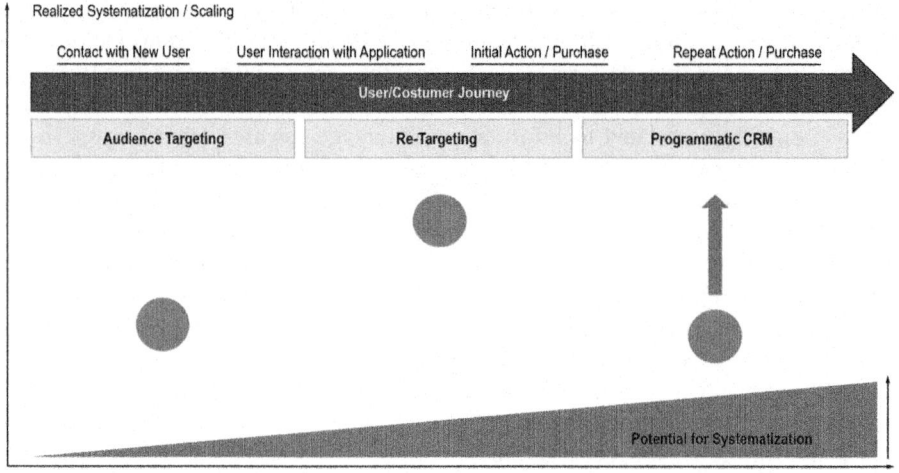

Fig. 1 Systematization/scaling

potential to systemically employ customer data in a scalable way and (b) the degree to which this potential is already used (Fig. 1), it becomes clear that CRM activities using programmatic advertising are clearly underdeveloped, even though the level of customer engagement is the strongest in the lower funnel which would allow for diverse angles to address customers.

2.2 Level of Systematization/Scalability

Very likely, retargeting offers the best structural possibilities for systematization, as specific customer behavior suggests clear triggers, based on which personalized promotional activities can be initiated. And in this field, due to players such as Criteo, Sociomantic, but increasingly also Google and Facebook, the existing potential is already exploited to a large degree. In a way, retargeting is a precursor of programmatic CRM and the lines are blurry between the two.

Systematic use and scalability of programmatic prove to be much more difficult to achieve at the beginning of the user/customer journey, the so-called "upper funnel". With regard to audience targeting, experience thus far has shown that there are only few data points or types of input data apart from current search referrer data that really predict in a reliable and scalable way whether a certain user will convert for a specific product or service. This appears to be a structural problem. As a result, apart from the "high/deep reach login platforms", there are few players that focus on and really scale acquisition of new customers in programmatic advertising. Rather, in order to demonstrate seemingly attractive levels of performance, retargeting campaign elements are often mixed with audience targeting/upper funnel activities. This combination, on average, leads to an acceptable performance. But as "de-averaging" and "granularity" are key principles of performance generation in advertising, this path is not very likely to be sustainably successful. As an exception, only the "high/deep reach login platforms" based on their superior data have a fair chance to build performance prediction models that scale by combining multiple data points.

In CRM, advertising messages are displayed to existing customers based on customer segmentation, which is often related to or derived from the purchase history of a specific customer. The main channel for digital CRM is still email. Mobile push notifications and Facebook custom audience are rapidly gaining importance, though. But there is a lot of – thus far largely unused – potential for programmatic advertising here. Especially as there is a real opportunity to generate a competitive advantage based on the asymmetric information inherent in first-party CRM data. Moreover, customer segmentation data by definition have a longer lifetime than the behavioral data that are the basis for retargeting and the models that operate with these data are very hard to re-engineer from the outside. This is particularly true if an advertiser is not only using email, push notifications, Facebook custom audience, and programmatic in an isolated, "silo-by-silo" way but rather in an orchestrated approach acknowledging interaction effects.

Implementing and continuously developing this approach in a sophisticated way allows an advertiser to build a sustainable competitive advantage compared to other advertisers addressing a similar target group. This also holds true for startups or other new entrants in a market, since a well-orchestrated CRM approach implemented from the start can significantly strengthen a business model and generate the necessary boost to push a company into structural profitability. In this case, structural profitability is defined as a clear path to profitability, i.e. profitability on a per customer basis with a realistically reachable number of customers in sight that would enable overall operative profitability of a model. In this regard, CRM functions as an essential complement to search engine optimization (SEO), which has previously often been utilized to lower the marketing cost ratio and enable structural profitability. And since SEO increasingly has become a mid- to long-term exercise, new entrant advertisers should put a strong focus on CRM earlier in their lifecycle.

2.3 Scaling the Approach

A key question, however, remains the scalability of the approach. While retargeting depends on users that already visited the website and/or app, the basis for CRM activities is even narrower: They target only existing customers. But a deep understanding of these existing customers builds a promising basis for further scaling of the approach. This follows the same logic as the lookalike campaigns connected to Facebook custom audience. Having mastered "core CRM" that is based on traditional customer segmentation and applying it to programmatic advertising should enable at least some further scaling through "lookalike CRM" and also "engagement CRM" that uses behavioral triggers, i.e. has a strong overlap with retargeting (Figs. 2 and 3).

In any case, it is clear that "engagement CRM" and "lookalike CRM" will produce inferior performance compared to "core CRM". The extent to which "core CRM" can be scaled depends heavily on (a) the quality and predictive

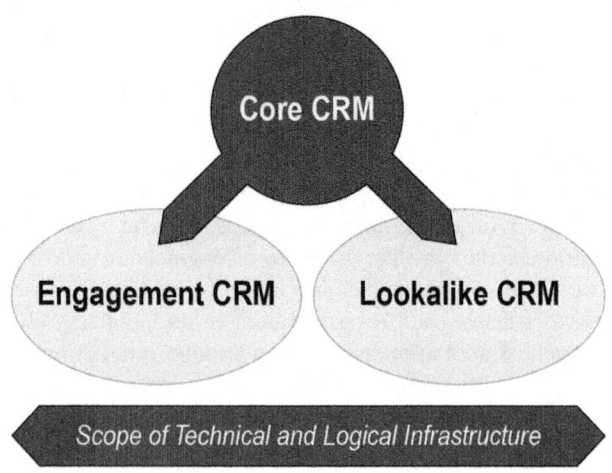

Fig. 2 Core, engagement, and lookalike CRM

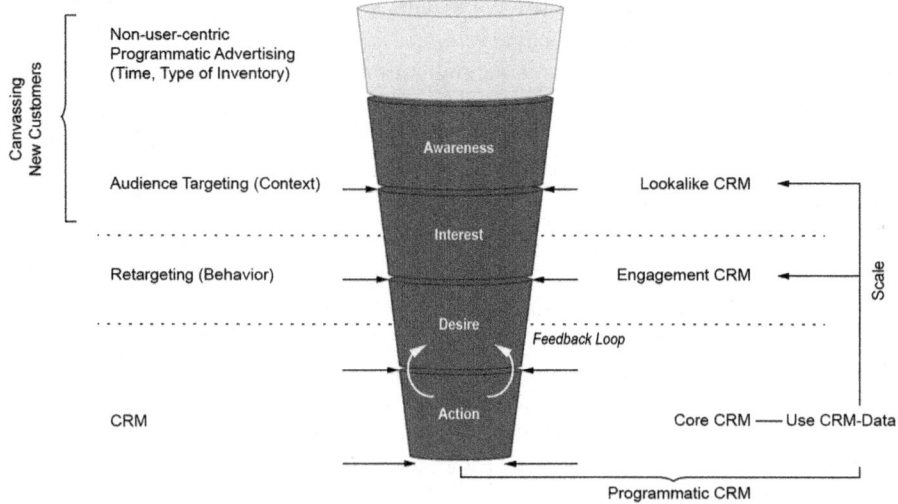

Fig. 3 Funnel

Driving Performance with Programmatic CRM

power of existing customer data and (b) the ability to generate and apply lookalike models. The latter are often already included in demand side platforms (DSPs) or provided by companies such as Semasio.

In any case, all three CRM-related approaches should ideally be operated on the same technical infrastructure. And moreover, there should be a reasonable level of orchestration across the various options to contact customers and derived users.

3 Developing an Orchestrated Approach

3.1 Taking into Account the Total Cost of Contact

Of course, not all CRM measures are composed equally in terms of the costs involved (Fig. 4). The cost of a specific contact has to be taken into account as one key dimension in designing the "right" orchestrated approach for a given user segment. The pure media cost should not be the only element taken into account: The costs for distribution and design/creation also have to be considered.

Using programmatic CRM to approach existing customers has a good chance to produce a strong performance KPI, but on the other hand is associated with relatively high costs. Therefore, it often makes sense to use this channel (a) primarily for high-quality customers and (b) if other, less expensive techniques failed to generate a response.

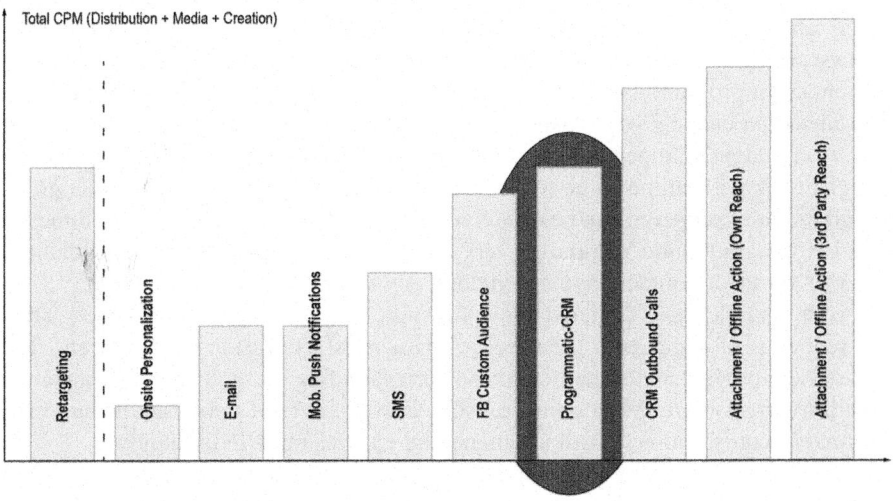

Fig. 4 Total CPM

3.2 Arriving at an Orchestrated Approach to Contact Existing Customers

In a best-case scenario, programmatic CRM is used in an integrated way alongside other measures/techniques. Such an orchestrated approach involving multiple methods requires that (a) all user/customer journey data are systematically collected on a per-user level and (b) are made available across all different channels. For most advertisers this is currently not the case, since they, at least regarding some channels, rely on third-party tools that do not allow to access such data on user level, i.e. those tools constitute a "black box" for advertisers.

Moreover, an orchestrated approach requires an infrastructure that combines all relevant components in one integrated tech stack. It is good news that today all required components can be sourced from third party tool providers, so that advertisers can really focus on initially defining and continuously improving the logic of their cross-channel CRM approach.

The central component of this orchestrating system therefore is the logic and reporting engine in which the underlying models are implemented. Based on the individual user and user segment information they determine (a) which promotional material is placed (b) when (c) how/in which format, at (d) what cost, and (e) in which frequency in order to produce the best return on marketing/CRM investment.

3.3 Example Cases

As mentioned above, ideally, programmatic CRM measures are part of a coordinated approach to contact users/customers. Since one, as part of the bidding process, directly competes with retargeting bidders and brand advertisers, the cost of contact in programmatic CRM tends to be relatively high. One therefore has to consider on a case-by-case basis to what extent programmatic CRM makes sense and what budget to allocate to this channel. For example, in case of customers with a high customer lifetime value (CLV), starting the contact sequence with a high bid programmatic campaign can be effective. In case of lower monetizing customers, a low-bid programmatic campaign very likely makes sense later in the chain if cheaper forms of contact have not yielded any response (Fig. 5).

As the sample cases depicted above illustrate, it very likely makes sense to adapt the sequence of contacts to the specific characteristics of each user segment. This cross-channel view and in particular the corresponding cross-channel management is still relatively rare. Most companies/advertisers, even if they use attribution on customer acquisition activities are managing each channel individually.

This is manifested by the fact that in most companies, the responsibility for different channels lies with different people or teams within the marketing organization, e.g. the display team would be responsible for programmatic display and often also handles Facebook advertising, the mobile marketing team is in charge of push notifications, and the CRM team takes care of emails. All teams, though, are

Driving Performance with Programmatic CRM

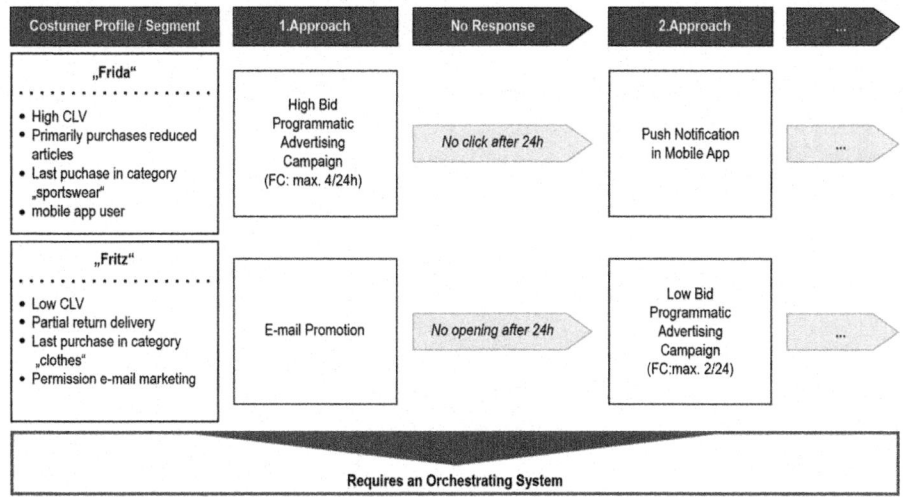

Fig. 5 Example cases of contact sequences

targeting the same customers but completely leave interaction and effects of sequence out of the equation. That is why an orchestrated management is so essential.

3.4 Infrastructure

The logic and reporting engine forms the central element of the tech stack/system. It is fed with data aggregated from different sources. And based on this data, models that determine which sequence of contacts to apply to which customer segment are implemented (Fig. 6).

Sources of Data/Input Typical data sources include behavioral user data (including cross-device), transactional data, CRM data, and potentially also third-party data, e.g. from players such as Eyeota, to complete the picture. While the latter is often too expensive to be used in the context of performance marketing, they may very well work for CRM purposes due to the typically higher response rates on CRM messages.

Logic and Reporting Engines In the context of programmatic CRM, the central logic and reporting engine has to ensure the coordinated usage of the data management platform (DMP), the bidder/demand side platform (DSP), the ad server, and the creative. It also typically controls the connection to third-party systems/tools via an API, contains an optimization layer, and a reporting engine. In order to successfully implement programmatic CRM measures, the appropriate CRM data points on existing customers have to be made available in the DMP. The optimization layer

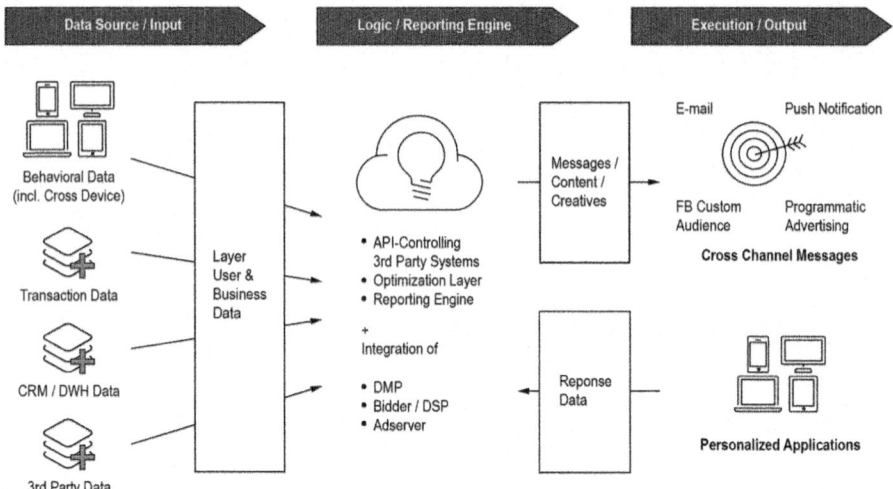

Fig. 6 Infrastructure

determining the campaign logic is probably the most critical piece of the infrastructure. In the end, the quality of this layer is directly linked to the extent to which CRM can be sublimated into a competitive advantage.

Execution/Output Ideally, all channels/points of contact with an existing user are managed based on the very same infrastructure. So, irrespective whether a user receives an email, a push notification, a programmatic CRM banner, sees an ad via Facebook custom audience, or a personalized message on a website – all measures should be derived from the same coherent logic. And the response data generated from the different activities can in turn be used to optimize the models within the logic and reporting engine.

4 Comments on Specific Implementation

4.1 Source of Data: Problems with Legacy Systems

The availability of user and business data in a suitable format is a key precondition for the aggregation and utilization of user/customer data and thereby for the realization of the above described approach. For startups or other new entrants in a market, this is not a major issue, since they are able to apply proper data preparation right from the start. More established companies, on the other hand, may face severe obstacles due to data spread across a variety of legacy systems. This complicates to merge these data in a single data source. Therefore, it is essential to address these kinds of problems beforehand.

4.2 Operational Suspension

While implementing an orchestrated multi-channel CRM, its concept should be mirrored in the operational structure. A cross-departmental operational suspension is required to conflate and distribute information and experiences across all involved teams. A cross-functional team is certainly useful in this respect.

4.3 Infrastructure: Third-Party Components Versus Self-build

Existing infrastructure and tools should be favored, if possible. Resources should be utilized to create additional value, and tools should only be self-build when they enhance differentiation. Most attention should be paid to the logic and reporting engine since the quality of the entire system is highly dependent on the underlying models and concepts. As long as there are alternatives, rebuilding the infrastructure is, if at all, only worthwhile for larger companies. Smaller and mid-sized companies should prioritize other aspects instead.

4.4 Optimization: Multi-channel Marketing Increases the Complexity of Channel Orchestration

Giving careful consideration to the optimization layer is crucial as well. Due to the fact that there are several channels that need to be orchestrated and optimized in form of contact sequences, response optimization proves to be significantly more complex compared to the, more or less, isolated optimization of single channels. Since each channel already contains a number of possibilities and these in turn have to be combined in a variety of steps with other channels, the complexity of the whole system that needs to be optimized is increased significantly.

4.5 Direct Versus Total Response

One aspect of general importance which should be stressed in this context is the distinction and correlation of direct and total response. The latter should by all means be taken into account, as a high direct response rate may initially appear to indicate high performance, while the actual total response rate might turn out to be much lower than suggested by the direct response measurement. An example would be "fake ads" that have a good direct response, but annoy the customer and lead to a poor total response. Of course, this discrepancy can also occur in an extenuated form, without intentionally placing misleading ads. Proper models to adequately measure total response will enhance optimization. But in general, with the little attention one can expect from users, it is crucial to provide them with a positive experience.

5 Conclusion

Utilizing programmatic advertising in an integrated and comprehensive multi-channel marketing strategy based on CRM data is a promising approach regarding its potential of differentiation, systematic scaling, and thereby of ensuring a competitive advantage in the future. For that reason, programmatic CRM measures should be integrated within a well-coordinated approach to contact customers as well as potential customers. If the current trend continues and "high/deep reach login platforms" continue to gain importance, advertisers will have no choice but to face the problems described above.

In this case, each transactional business model in a competitive market will have to implement such a system in order to remain competitive. Established brands might be an exception in this respect. But they, too, can benefit from a systematic use of programmatic CRM in a coordinated approach of addressing customers

Since the quality of the orchestrated system is primarily measured by the underlying models and only up to a certain extent by the amount of available data, first-party CRM data does suffice to generate actionable insights that are nearly impossible for outsiders to replicate.

Dr. Florian Heinemann is a Founding Partner of Project A Ventures, a Berlin- and São Paolo-based early stage investor and company builder. He is responsible to develop Project A's platform in the areas of marketing, CRM, and business intelligence (BI). Before starting Project A, he was Managing Director at Rocket Internet holding the same responsibilities (2007–2012). During his time at Rocket, he was deeply involved in companies such as Zalando, Global Fashion Group, and eDarling/Affinitas. Prior to Rocket, Florian was Co-Founder/Managing Director of JustBooks/AbeBooks being in charge of marketing and product (1999–2002, exit to Amazon). Then, he was co-heading the performance marketing department of Jamba! and the online dating portal iLove (2003–2005, exit to Verisign). In 2006, he was Co-Founder of the leading digital marketplace research antibodies, antibodies-online.com.

Moreover, he has been business angel in more than 80 startups, e.g. AdScale, Audibene, Ladenzeile/Visual Meta, Tradoria, and Trivago. Florian holds a doctorate's degree in innovation management/entrepreneurship from Aachen University (RWTH). He was a visiting scholar at the Wharton School. His research has been published in leading international journals, such as Strategic Management Journal, Journal of Product Innovation Management, Journal of International Marketing, and Zeitschrift für Betriebswirtschaft. Furthermore, Florian holds a masters degree in business administration from WHU Koblenz. During his studies, he was a fellow of Studienstiftung des Deutschen Volkes (German National Academic Foundation).

Pricing for Publisher: Scaling Value, Not Volume

Marco Klimkeit and Paul Benson

Programmatic Advertising is frequently considered as being synonymous with remnant inventory advertising. In fact, in many markets there are cases in which so-called premium publishers sell their inventory through Programmatic Advertising or, at least, make parts of it available to advertisers and their media buyers. Premium publishing and Programmatic Advertising do not contradict one another. This chapter explains how publishers of premium inventory can profit from the programmatic media trade. It became clear that these publishers even increased their revenue by using the right pricing strategy. Therefore, direct selling will continue to be a strong revenue tool in the era of Programmatic Advertising. The focus of this chapter will be on existing pricing strategies of advertising media and the most important factors that can influence the price of an ad impression. The chapter will also discuss the common arguments against Programmatic Advertising.

1   What Is Premium?

Premium is the special advertising effect of a brand environment in combination with the advertising format and the audience. Premium includes not only brand-safe inventory but also a quality environment that can have a great impact on the audience; it can "make the mental click", so to speak. Premium is the inventory of the advertising media where the brand advertiser can reach his campaign goals.

M. Klimkeit (✉)
Yieldlab AG, Colonnaden 41, 20354 Hamburg, Germany
e-mail: m.klimkeit@yieldlab.de; mail@marcoklimkeit.com

P. Benson
Adition UK, 5, St John's Lane, London EC1M 4BH, UK

It is well known that brand advertising in the environment of popular media brands, such as Spiegel Online or heise.de, works much better than in so-called long-tail inventory. The reasons for this lie primarily with the media themselves, which have developed their media brand for decades. Sophisticated and professionally generated content communicates respectability, credibility and sustainability among a user group that to a large extent has a higher level of education and an above-average purchasing power; these quality inventories provide advertising with the perfect mix of brand safety, premium audiences and appeal for their own brand.

The great demand in the form of bidding volume and the bid prices achieved in quality inventory prove that Programmatic Advertising works particularly well for premium publishers. The bid prices are considerably higher than for a long-tail inventory. This is because real premium inventory is available only to a limited extent and therefore remains a precious and sought-after asset for the advertiser. The automated selling through supply side platforms does not change these principles. On the contrary: Programmatic Advertising offers publishers of premium inventory an additional powerful tool where they are able to channel and control the demand and bids of the buyers by means of different pricing strategies in a targeted and granular fashion.

Programmatic Advertising leads to determining the "true value" of one's own inventory. Price developments show that advertisers are willing to offer a higher bid for ad impressions on websites with quality content than for advertising contacts in the mid or long-tail.

The topic of brand safety for the advertiser thereby increases. This becomes obvious to a supply-side-platform (SSP) provider because a growing number of advertisers deliberately submit their bids to transparent URLs, for example. For their campaigns, brand advertisers do not want to rely only on cookie data. The exclusive "dependence on cookies" is a misconception from the early days of the programmatic movement. The media brands and their publishers are on a learning curve. After they used their SSPs to release only their remnant inventory for Programmatic Advertising, this changed with the increase in demand. In the meantime it has become the custom of quality publishers to offer high-quality inventory classes in a transparent way through their own SSP to achieve higher advertising revenues and yields.

In order to ensure that this happens, it is absolutely crucial for Programmatic Advertising that the supply side remains in control and does not offer the inventory haphazardly as is the case with discount wars on Demand Side Platforms (DSPs). Supply Side Platforms allow you to see how strongly the number of ad spaces on offer relates to the resulting bid price. The higher the number of ad spaces offered to the advertiser, the easier and quicker it is for him to reach his users with a low CPM. That is why media and their publishers should bank on technologies and Programmatic Advertising systems that allow them not only to remain in full control over their own inventory, but also and above all, to use their own pricing strategies that match the different inventory classes perfectly.

2 The Right Programmatic Advertising Strategy Is Key

Many premium publishers still have misgivings about the potentially negative effect of Programmatic Advertising on their own prices. This prejudice mainly stems from experiences with the purely auction-based realtime bidding (RTB), which prevailed in the US market especially and today can be found in many other developed regions only for long-tail inventory.

In fact, we can observe the completely opposite development: The right Programmatic Advertising strategy helps attractive web offers to gain traction. After all, it is up to the medium or its publisher to establish rules for the publishing of each individual ad space. This can occur on several levels: Publishers can publish parts of the inventory through a traditional yield optimisation or set floor prices on the level of the individual ad space or each individual website. Depending on which data are supposed to be made transparent to the demand side – such as the entire URL or higher-ranking parts only – the supply side is able to decide on private marketplaces, for example, which advertisers are allowed to participate in an auction of specific inventory classes and which are not. This way, the supply side remains in control at all times. New clients can then be won, risks minimised and higher revenues generated because the cannibalisation of the sales to direct clients is precluded.

2.1 Pre-Negotiated Deals (Deal IDs)

A core component of a modern marketing strategy of premium inventory is the Deal ID. Deal IDs are among the most important developments within Programmatic Advertising. Deal IDs are rightly considered as the strongest concept in today's online advertising. They help to transfer pre-negotiated deals of direct sales or framework agreements with agency networks to a platform-based marketing without any difficulty.

Deal IDs are the link between buyer and seller. First, the publisher and buyer agree offline on the terms and conditions including price and ad space of the Deal ID. The result of these preliminary negotiations is then transferred to the SSP. The SSP sends the buyer a Deal ID, which provides access to all rules. In this way the buyer is able to gain direct and, depending on the agreement, privileged access to the inventory of the medium.

Unlike true auction-based realtime bidding, the Deal ID is used to determine together with the advertiser at what price he can purchase the ad impression rule-based via his DSP from the publisher. However – and it is very important to understand this – a predefined price with an advertiser does not necessarily exclude a bid on an impression. Why should it. The sell side aims at achieving the best price and yield for a page impression in a ranking. If another buyer outbids the predefined floor price, he will win the bid. The publisher achieves higher revenue with this bid. And that is what the publisher is primarily concerned with: the increase in his own advertising revenues.

As an option, Deal IDs provide the ability to grant an individual advertiser an exclusive look at an ad impression. This "first look function" gives select buyers the exclusive opportunity to submit a bid on an ad impression before anyone else. If this "first look" does not result in a sale, the ad impression is offered at the general auction. The SSP programmatically prioritises the client's Deal ID. The buyer with the highest priority gets the "first look". To determine the right floor price of these "exclusive offers", quality criteria and data influence the price and plays an important role for the publisher.

3 Price Factors and Quality Criteria

Pricing in Programmatic Advertising can be set by advertisers, interested buyers (e.g. media agencies) as well as suppliers and publishers. Depending on whether it exclusively relates to the pure auction process – that is the realtime bidding – or other pricing models need to be taken into account, a large number of data, factors and quality criteria can influence the value of an ad impression and therefore the pricing. A typical example of these factors is, for instance, the incorporation of targeting data or market research data obtained by the publisher. Generally, the following factors can influence the price of an ad impression.

User Data (On the Demand and Supply Side)
- Behaviour during campaigns/on websites (re-targeting information)
- External data sources
- Previous contacts with the campaign

Data of the Ad Space Offered (On the Demand and Supply Side)
- Data such as click rate, conversion rate, previous effective cost per thousand impression (eCpM)
- Own environment evaluation, manual or automated by means of URL analysis

Campaign Goal (On the Demand Side)
- Cost per activity (CpX)
- Term
- Goal of impression
- Performance/Branding

4 Pricing Strategies for Publishers

The system options of the participating platforms, the SSPs and DSPs, offer unprecedented opportunities for controlling and protecting one's own inventory. Of course, the publisher also has a price in mind for his ad impressions. That is why different rules are established on that side. Media and their sales houses can use the following pricing and optimisation models through Programmatic Advertising with the support of their Supply Side Platform. It is important to understand that these models can and should be combined with the Deal ID concept and with each other.

- Fixed pricing (without auction)
- First price auction (with RTB)
- Second price auction (with RTB)
- Dynamic floor pricing (with RTB)

4.1 Fixed Pricing

On different levels of the inventory classes, the publishers can, of course, define fixed prices on an SSP. The fixed pricing model is not actually a pricing strategy. Rather, the SSP uses the predefined CPM prices of the marketing unit. However, in this case the publisher still enjoys the typical efficiency benefits in the automated media trade, without the inventory that has been priced in this way participating in the auction, or the use of further yield and pricing models. However, Programmatic Advertising only reaches its full potential in an auction-based process rather than in fixed pricing.

4.2 First Price Auction

In addition, the media owner can use two different auction models for selling his inventory. For higher priced inventory classes, it is recommended to use the first price auction. This auction model is particularly suitable for inventory with high CPM targets because the bid submitted determines the price of the ad impression. However, this also constitutes a considerable disadvantage of the first price auction, which should not be omitted. The number of advertisers on the demand side who participate in a first price auction is smaller and they tend to avoid this auction model for financial reasons. Still, a first price auction for higher priced inventory classes can make sense and provide a higher yield in the long term than the widely used second price auction.

4.3 Second Price Auction

The bid of the buyer determines the price in a second price auction. In this type of auction, the highest bidder has to pay only the second highest bid plus 1 cent. Empirically, this auction model is more suitable for inventory classes with lower floor prices, which are intended for the largest possible number of bidders. The average CPM evens out very dynamically. The second price auction is the fairest auction model unless the publisher optimises with dynamic floor pricing at the same time.

4.4 Dynamic Floor Pricing

Dynamic floor pricing is an optimisation opportunity where the SSP adapts the floor price based on the demand. With this optimisation model, the SSP analyses the demand and the floor price for the ad space according to the highest bid from the past and dynamically adapts the floor price during the auction. Therefore, if a particularly high number of advertisers submit high bids on an ad space the floor price rises accordingly.

Dynamic floor pricing is a generally accepted process in the market; however, it rightly meets with much criticism because it turns the bidding process into a reductio ad absurdum. This is because the current floor price in an auction is based on the analysis of previous auction processes rather than the actual demand in the course of the auction. Dynamic floor pricing on the basis of the bidding behaviour fails to provide the necessary transparency for advertisers and can even lead to the demand side avoiding the bidding model altogether in the long term.

5 Interim Conclusion

Overall, a premium publisher benefits from using a system to control and optimise the campaign of both brand- and performance-oriented demand.

> ▶ **Only a centralised system allows the publisher to react quickly to any changes in demand in the individual pricing segments and to adapt his own pricing strategy to these changes.**

In future, we will be able to observe a closer connection between marketing systems (CRM, DMPs, booking systems) and the SSPs. As a result, the floor prices will be matched automatically to the activities of the direct sales. (Fig. 1)

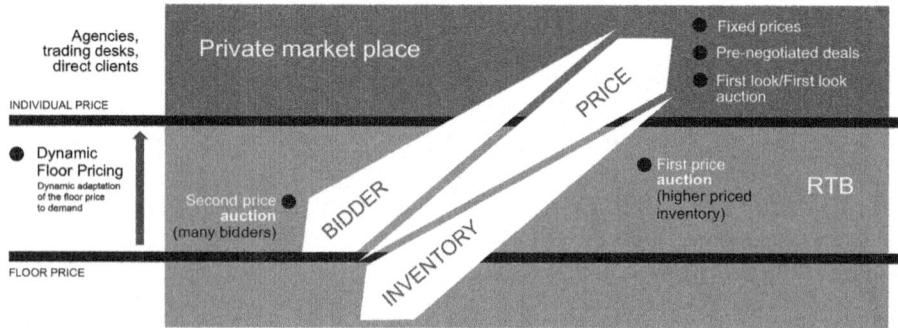

Fig. 1 Utilising the full potential of a centralised system

6   Added Value Through Efficiency

Automated processes in the sale and purchase of online media provide clear benefits: In the traditional display advertising many processes between publisher and media agency run "analogue". It is still common that the contracting parties exchange emails about offers, booking orders, ad tags and reports. If a problem occurs, for example a difference in numbers, it has to be solved over the phone. Programmatic Advertising digitises these processes, corrects current system errors and allows both the publisher and the buyer to focus on what is most important.

Besides these improvements in efficiency, advertisers are provided with extensive options to purchase media volume efficiently. This makes it possible to select the suitable advertising contacts from a wide range of publishers and to control a campaign by means of frequency capping, taking into account the suitable advertising impact.

7   Price Trends and Performance Records

For the publisher, however, the automated media selling also has further benefits. Through their SSP, publishers are able to identify quickly those inventory classes that are in unexpectedly high demand. It is the eCPM, that gives the publisher in an SSP information about the relation between campaigns and bidding prices. Through this performance, those rankings that work in the eyes of the performance-oriented advertisers can be identified with precision.

A higher floor price can then be set for these best performers. Ad spaces that are not in great demand can be taken out of the inventory of the publisher. The result is a reduction in ad spaces, which increases the overall quality of the inventory because less advertising is displayed.

The number of bids on the individual ad impressions may vary greatly. Regularly, there are ad impressions that do not attract any bids at all, while others attract 30 or more. A true bidding competition with an average of more than two or three bids above the floor price set by the publisher primarily takes place in the medium and low pricing segment. Therefore, the premium price can be achieved the quickest and most definite in these price segments. As mentioned above, premium inventory is normally offered via Deal IDs between the publisher and demand side. For optimising the brand-oriented campaigns, the same standards apply as for traditionally booked online campaigns with the important exception that through their SSP the publisher is able to control these campaigns in their environment in a much more granular fashion.

8 Client Trends

The platform-based media purchase allows media access to a completely new client segment, which they used to be able to cover only via ad networks – if at all. This is has been made possible because established SSPs are linked with a number of international DSPs and trading desks of agency groups. In the FMCGs (Fast Moving Consumer Goods) sector especially, agency holdings have started to use Programmatic Advertising to purchase their global branding campaigns via trading desks across borders from premium publishers. DSPs or trading desks are assigned their own "trading seats" at the big ad exchanges and SSPs in order to place and pay for the campaigns directly.

The platform-based media trade provides publishers new and profitable clients also from another side. Programmatic Advertising facilitates the direct access to high-quality inventory and interesting audiences. Many businesses and brands have therefore started to place their advertisements directly with the publishers, thus avoiding media agencies. That is why an increasing number of advertisers will have their own "seat" in the SSPs in future.

9 Common Arguments Against Programmatic Advertising for Premium Publishers

- **Programmatic Advertising leads to a loss of control and cannibalises one's own publishing business:**
  The strongest arguments against Programmatic Advertising are a potential cannibalisation of the direct sales and a loss of control over one's own user data. In the premium segment of the display market especially, user data is an important asset of the portfolio which has to be protected. The current technological possibilities for the media purchase on a single user basis enable all connected buyers to eavesdrop on the inventory on a large scale, with many implications. Unfortunately, there are hardly any technical barriers to prevent this data leakage. This does not, however, apply to the loss of control when selling the inventory. The publisher can protect his existing campaigns and revenues through whitelisting and blacklisting. This ensures that an advertising client is unable to buy an ad impression at a lower price through Programmatic Advertising than through a top price campaign run in parallel. Deal IDs also prevent a cannibalisation of the direct sales.

- **Programmatic Advertising is not suitable for branding formats:**
  Another long-standing objection was the view that both Programmatic Advertising and realtime bidding is only possible for standard advertising media. This may have been true back in the early days of programmatic advertising, but in the meantime this objection has become obsolete. Many Programmatic Advertising campaigns, which are also bought auction based, in the meantime bank on diverse forms of branding advertising. One example out of the premium-driven

German market: AdAudience, a joint venture of leading premium publishers, succefully offers special formats per Programmatic Advertising through the Yieldlab SSP.

- **Programmatic Advertising is expensive:**
Another objection relates to the costs for the integration and the use of a Supply Side Platform. It is often presumed that the gain in efficiency through Programmatic Advertising is minimised by the high costs for the technology. It is correct that the establishment of a Supply Side Platform is complex and costly. It is important, however, to differentiate between the different SSP models available on the market. Most platforms reduce the investment risk for the publisher enormously because no or very low fixed costs are incurred and accounts are settled through profit sharing. Besides, the systems can be booked either as a self-service or as a managed service. With the self-service, the existing yield management team of the publisher is able to manage the platform. The managed service has the advantage that no further personnel resources are necessary. Some platforms recommend forming one's own in-house team consisting of several members who manage the platform. This may lead to a high investment volume for customising and support. The publisher should thus consider his total calculation very carefully.

- **Programmatic Advertising is remnant inventory advertising:**
The pure so-called open RTB model is on the decline in the US as well as in Europe and other regions, with excpetion of the Google universe and Facebook as well as a few smaller long-tail markets. For a long time the models have been shaped by US companies without taking into account the actual requirements of less fragmented, premium-driven markets. It can therefore be expected that those markets the media selling of premium inventory with the help of Programmatic Advertising will increase considerably over the mid- and long term.

10 Conclusion

Programmatic Advertising has got nothing to do with remnant inventory. Rather, it combines the opportunities of a user-based target group system with an automated media purchase. Programmatic Advertising connects advertisers and their agencies with the media houses and their publishers – in performance businesses and for premium inventory. The first SSP providers have found ways to tackle the topic of Programmatic Advertising along the circumstances of the premium-driven markets. Every publisher needs his own pricing strategy that matches his requirements and an individual implementation that provides him with complete control over his own inventory even in the era of the automated media trade.

Marco Klimkeit has been working in the online industry for more than 14 years. In 1999 he founded a technology company in the ad serving sector which was the first company in Germany to offer retargeting. In 2010 Klimkeit founded the first German Supply Side Platform, Yieldlab, which is used by many european media houses. Klimkeit lives with his family in Hamburg.

Paul Benson has worked in the online advertising industry since 2000. He has spent most of his career in the ad operations space working for large organisations in the UK and abroad. In 2010 Paul founded Adopstar, an ad operations outsourcing company, which provides managed ad operations services including programmatic advertising. In 2015, Paul co-founded ADITION UK, an online advertising technology company providing Yieldlab SSP, the ADITION ad server and Mov.Ad, a rich media and video ad-building platform.

Managing to Quality Attention and Outcome Through Programmatic Technology

Ted McConnell and Lothar Hoecker

With most online media, the compounding effects of low attention quality have enormous impact on advertising ROI. Contributors to low attention quality often include "opportunity to see" the ad; video completion rates; out-of-geography impressions; ad load success; over frequency of impressions served; non-human traffic; and target accuracy. Further, attention quality is, in most cases, measurable, and, in most cases, manageable. These factors create a case for managing to attention quality; the benefits are clear, immediate, significant and certain. Programmatic technology allows advertisers to better solve these fundamental questions in media and communications. We suggest a shift away from pure focus on CPM and stronger leverage of optimization through programmatic technology. Yet – technology cost need to be monitored closely with reaching up to 50 % transaction value today and Google and Facebook aiming to dominate this field.

1   Quality Attention Drives Outcomes

It's obvious that if an ad is not seen, it can't do any good (with video sound, or audio content, its opportunity to hear). However, attention is not a binary quantity; there are degrees of attention. From a measurement standpoint, attention is a phenomenon that occurs in the mind of the viewer and cannot be measured directly in scale (i.e., brain wave measurement), so, we infer attention from dozens of available

---

T. McConnell (✉)
Ted McConnell Consulting Llc, Cincinnati, Ohio, USA
e-mail: ted.mcconnell@me.com

L. Hoecker
Dentsu Aegis Network Germany, Wiesbaden, Germany
e-mail: lothar.hoecker@dentsuaegis.com

signals online, including the likelihood that a viewer wanted to pay attention ... their receptivity. As in most media measurement, perfection is unattainable, so we live with surrogates.

Many offline ads get attention in part because an ad, for a moment, is the only thing in view, but in the web, each ad competes with half a dozen objects vying for the user's attention. As a result, attention on the web is more difficult to get. However, unlike TV, attention quality on the web is *less* difficult to measure. Further, attention metrics can be put into action readily, mid campaign.

Meaningful attention is more than a synaptic response to sensory input. To be useful to an advertiser, attention must be from humans and not robots; it must occur for long enough to get the message; it must be from someone who might plausibly purchase the brand, and so on. Anything else is waste.

Attention drives audience quality and effectiveness, and attention does not happen without the opportunity to see.

2 Measurement: Managing to Impression Quality Versus Brand Impact

Central to the decision process of any advertiser is the question of whether the advertising is worth it.

This article provides a simple a calculus for estimating the cost of waste, ways and means to avoid it, and to opportunities presented by reducing it. We will address specific, situations that result in waste in hopes of helping planners and buyers focus on strategies that will be effective.

Why focus on waste instead of the usual questions regarding value?

Typically, efforts to measure success are designed to try to attribute exposures to a change in consumer behavior. Someone saw the ad, and then they bought the product, or did not. But did the exposure cause the sale?

Indeed, CMO's and media managers want to know whether advertising caused sales, and so they should. With direct response advertising, it's pretty simple. But with pull through, or indirect marketing, its much harder. Here, it's important to remember that causing marketing ROI to go up is the real goal, so there are two paths to success: make the return go up, or make the investment go down. Eliminating waste makes the investment go down.

We are suggesting that controlling and measuring what didn't have an impact might be a faster route to ROI than estimating advertising response to indirect sales.

The reason is pragmatic: it's hard to tell what worked but it's relatively easy told what didn't. This works especially well online because impressions that didn't help are often a huge portion of the total.

Brand sales-based measures suffer uniformly for lack of control for co-linearity, sample bias, and more. For example, packaged goods companies launch in all marketing channels simultaneously: media, promotion, coupons, discounts, retail feature, etc. given this, it's hard to reliably measure which channels were causal in the purchase decision.

More compelling, impact measures, because of time and aggregation effects, are difficult bring into mid campaign action. They arrive too late, or are not actionable because results are aggregated differently than the buying decisions.

On the other hand, waste oriented measures can feed back to programmatic advertising optimization in real time, and if not, be used to adjust campaign parameters such as white lists.

Another persistent question our model will provoke is whether it's fair to compare waste among different media. Most media buyers believe that some impressions will have more impact than others for the same level of engagement. Research validates this. Reasons for this include aspects of receptivity, related emotional impact, time in view, temporal proximity to the purchase opportunity, etc. These are important factors, but in most cases difficult to validate.

We will identify and quantify the impact of useless, or wasted media, and note the currently available means to measure, and correct for it. Assuming waste percentages nominally between 80 % and 95 % are nominal for digital display, managing our media execution towards true exposure with audiences that matter has more potential to improve advertising ROI than other aspects of planning.

It's an uphill battle to reduce waste, but a battle with a clear and certain outcome. It's also an uphill battle to calculate impact, but that is a battle that's been fought for years but remains a subject of intense debate. The pursuit of "productive media" (1/waste) should be a prerequisite to the battle to measure impact. With respect to attribution, waste = noise. With 90 % less noise in our ROI models, we will have a much better chance to win the battle to attribute impact. No one will argue that eliminating noise in an attribution model is wasted effort, and clearly, waste is noise.

To do this, we need to start with identifying *sources of waste* that can be measured and at least partially corrected. After that, we will propose a simple model that shows advertisers how to calculate the potential ROI gain from eliminating various types of waste and finally propose on outlook on what advertisers should do.

3 A Note on Reach

Media mix models do not usually take reach into account. Uniformly, a smart media plan gets the high ROI first. Then as reach increases, ROI usually declines. Some aspects of waste play off against relevance (on target, for example). Our model does not take reach into account. It simply says that, if, in aggregate, your impairments are a certain portion of purchased impressions, then the effective cpm is x, across a mix of impairments (waste). This is a simple model anyone can understand. Eliminating impressions that have an ROI of zero makes ROI go up.

4 Sources of Advertising Waste

4.1 No Opportunity to See (OTS) ... Or "Not in View"

In Television, OTS is the standard for exposure. If an ad did not air, or, no one was in the room, there was no "opportunity to see" the ad. Online, if an ad never appeared on the screen, there was no OTS. Online, the new viewability standard is an OTS concept. To be clear, OTS is only the baseline for exposure. OTS *suggests* that an ad was seen, but is no guarantee of that. Think as OTS as a prerequisite for any engagement, measurable for TV (with limits) and Online (almost perfectly), and a surrogate for exposure. We need to go beyond OTS to get an idea of meaningful attention. What we want is quality attention from the right people, and OTS measures do not provide this, but do provide the first filter.

To go beyond OTS/Viewability towards real exposure, the web offers a range of signals from browsers that can yield excellent indications of attention. One measurement standard is "UIC active hover", an IAB standard. Certain publishing formats almost virtually guarantee OTS ... formats where there was no competition for attention. UIC active hover, according to Moat research, provides almost perfect correlation to Awareness in Brand lift studies, and is available inexpensively, and in census, making it vastly superior and more usable than traditional Brand Lift intercept surveys.

Over the past years there has been a significant development in measurement technology for viewability. And although there is not yet a formally approved "yardstick" similar to the "standard meter" in Paris, it is pivotal for advertisers and agencies to generate data and insights in this field.

TV delivers against objectives by delivering until an after-the-fact measurement threshold is met. Internet media does it the same way. Because viewability depends on the user's reaction to a page, the impression is bought before it is known whether a user will scroll. Viewability must be determined "post impression". Exchanges sometimes offer bid requests that include only a "probably of in-view", based on historical averages for the page (not even the ad frame). The buyer gets to pick a threshold, and take their chances. However, after-the-fact measurement of viewability is fairly accurate, so, like TV, reconciliation can be computed if the right number of viewable impressions is not delivered. With online media, that correction can be computed in seconds. And programmatic technology allows synthesizing the data collected into adjusted buying and planning strategies in Near-/Real-Time.

This also highlights the necessity of strong technical integration – while almost any AdServer today offers basic viewability measurement, it often lacks the necessary granularity allowing to deduce optimization strategies.

The nominal assumption in our Model is 40 % for not in view (60 % in-view), worse for exchange traded, and better for premium. These default inputs reflect well-established averages across several major types of supply including premium publisher and exchanges.

4.2 Video Completion

The value of video completion depends on the objective of the advertiser, and the symmetry of the creative in any media encounter, opportunity to see might have occurred, and the required duration or focus will vary.

Online, the IAB viewability standards call for 1 s for static ads, and 2 s of videos to "count" as having been in view. Note that online, it is safe to assume that someone who clicked on a video, or has invoked a screen, is paying attention to something on the screen, but you can't assume that with "lean back" media these days. In any case, in our model, an ad is treated as wasted if the completion is below some threshold.

Interest in video completion rate is attributable to the rise of online video, which has grown 20x in revenue since 2007 to a 13 billion dollar business in 2014. Consumers generally can x-out of any video at any time if they don't want to be bothered with it, and frequently do. Completion rates on TV are close to 100 %, and completion rates with online video are usually much lower, closer to 20 %. If the creative is a "reveal", the key message happens in the last few seconds. In that scenario, even 80 % completion means the ad was wasted.

Video might start with a select-in (click), or begin automatically when a page is rendered. This has major implications for OTS since generally, click-to-play display video is not subject to as much non-human traffic as auto play because benign bots don't generally click on video ads. Bots committing fraud are smart enough these days to hang around a site long enough to satisfy the video viewability standard, and Google's True View standard. With auto-play, of course, any execution of a page by a Browser (or Bot as Browser extension, etc.) will play the video. It might be sound off, of course, but it counts as playing, and therefore as an impression.

The model input is the percent of users who saw at least 20 % of a video. This is a low bar. In the nominal model, 70 % of viewers saw 20 % of the video.

Whether or not brand effects can be observed in that scenario is a creative opportunity.

4.3 Out of Geo

The web is global, and viewers cross international boundaries frequently and fluidly. In fact, it is estimated that as much as 30 % of all US traffic is from outside the US, yet many Brands in the US are sold only in the US. This will be less of a factor in countries where the native language is less ubiquitous.

An ad can easily land on a person/browser who could not purchase the product ... even if they wanted to. This is clearly wasted. The same phenomenon happens to some extent in most markets. Impressions such as this are wasted, and there are a lot of them in countries whose native language is spoken in many other countries.

Countermeasures are possible, but not as simple to implement or as reliable as one would hope. Publishers can detect the location of a browser rendering an ad from its IP address. This is somewhat, but not perfectly reliable due to corporate private networks, proxy servers, IP spoofing by fraudsters, etc. Even when they can correctly detect it, however, they can't do much about it.

The issue for publishers is that once the ad is rendering, even *if* they know the IP address and can respond to it inside the latency window, they have to find an ad that fits that geography. US publishers are unlikely to have creatives and decision infrastructure to quickly adapt to impressions coming from small countries or even big ones. It is possible for a website to simply block all access from outside the US, but this drives down traffic numbers so is not a first choice for many.

Larger platform sites such as Yahoo and Facebook often have declared geographies associated with browser information, and multinational sales ... so are much more likely to be able to respond to the sudden need for a geo-specific creative.

Other out-of-geo impairments occur with campaigns that are tightly geo targeted or even "geo fenced" in mobile. These can occur as a result of errors in geo information. One geo fencing supplier eliminates from their attribution data "any phone that appears to be going over 100,000 miles per hour"! Good choice.

In ten premium campaigns for US Brands measured by Comscore in 2008, about 20 % of impressions were delivered to places outside the US. In this study, all publishers had agreed (via agency contracts) *not* to serve non-US traffic, but most did anyway. So, 20 % waste is a good assumption for digital media except for walled gardens (i.e. Facebook) where a login, thus independent geo information, is provided.

In these cases it is likely to be close to zero. The problem has likely gotten worse not better since then, as both China and India have surpassed the US in internet connected users (thus US foreign traffic), but the basic motivations of the sell-side players have not changed much. Our nominal assumption in the waste model is 20 % ... or 80 % "in geo" except for platforms which are multinational and registration based, and TV. Again, the user of the model can input their assumptions.

4.4 Ad Not Loaded

Some ads just don't make it to a browser before the user presses the back key, or tabs out. Estimates from one Measurement Company peg this near 5 % of all impressions.

The nominal assumption in the model is 95 % ad loaded.

4.5 Over Frequency

From the standpoint of a buyer, every impression's value is dependent on its frequency. Not buying an over frequency impression, and spending the money instead on another impression, gets incremental reach. By either not buying too many frequency impressions, or converting frequency to reach, the effects of proper frequency management can dramatically improve ROI.

In the research community we get the same question about once a month: "What is the correct frequency?" There are several paradigms of frequency on which to predicate the answer, but there is no right answer. The right answer depends on how you think frequency functions in persuasion, learning, memory, and consumer behavior. Without going into detail, here are some of the dynamics.

One paradigm is that frequency functions as repetition. Repetition improves "learning". In most measurements, frequency improves recall scores in a declining curve called the frequency wear-out curve. After some point, additional frequency stops improving recall (the usual independent variable used to compute wear-out curves), so that point is the point at which its not worth buying more frequency. Often the curve flattens out between four and seven impressions. Note that exposure to the same message in a new medium usually extends the wear-out curve. There is also some evidence that a new context extends the wear-out curve. This is explained by the theory that our brain decides when to ignore something based how it fits with its context, so a new context causes us to reassess the message, while an old context causes us to filter it out.

Another paradigm is "one". Erwin Ephron in his classic treatise on Recency provides ample support for this idea: beyond the first real exposure, subsequent exposures function only to remind of us a brand. The idea is simple, once we get the story, subsequent impressions need not carry the burden of telling it again. A simpler message will do. David Ogilvy once quipped that the ideal frequency is One. He probably meant that better story telling and precision media execution are preferable to bombardment with mediocre creative, consistent with his idea that bad creative with high reach is more damaging than bad creative with no reach.

In the face of debate regarding whether the right number is one, or four, or one per week, or "7", per campaign etc., we have staggering measurements from modern web campaigns that show reality on the web makes those choices look academic. In *many* current campaigns online, something like 30 % of the impressions go to <1 % of the users. The only way to measure this is to tag entire campaigns, and count cookies. This technique usually shows massive portions of the audience getting one impression, and small portions getting in excess of 100,000. The typical advertiser may see an average of seven, and be satisfied, but in fact, almost no audience member is getting seven impressions. The users getting frequency of 100,000 are assumed to be robots, and averages mean nothing in a severely skewed distribution.

Since panelists are presumably not robots (although they can have bot infected browsers), panel-based measurements provide insight about frequency in the absence of non-human traffic. In an online campaign that calls for a frequency of four, the typical delivery might be double or triple that (to humans). The reasons for this are several. One is that users delete their cookies occasionally, or don't accept cookies, and therefore a delivery system cannot know if they saw the ad before. Another is that publishers and networks set their own cookie to control frequency, but are unable to see whether that same user has seen the same ad before in a different domain (different publisher). The third is that publishers handle campaign reviews periodically for flighting, and when they see they are behind, open the floodgates in order to meet a campaign impression goal. As a result, any user on that site, that day, may see a very high frequency during the course of a single session. The message here is that even in the absence of robotic traffic, frequency is difficult to control across a complex campaign.

Using programmatic advertising, the buyer can eliminate most of these causes with a single strategic choice: Flow all impressions via the same programmatic decision point.

Eliminating wasted frequency is a simple way to instantly double ROI ... but with current technology, requires a partnership with *one* distributor or network for all display media. This choice reduces the ability of a buyer to pit suppliers against each other, so buyers don't like to do it. However, the "commoditize and RFP" approach might get them a 10 % advantage, whereas the single control point approach is likely to get four times that. Since exchange pricing is generally transparent, it's not hard to get both advantages (pricing and frequency control) by awarding a single partner with all business.

Alternatively advertisers through their agencies can create Private Marketplace Deals which allow them to leverage the programmatic frequency capping/management opportunity while continuing to work with their selected publishers and prices.

How much? If you are managing frequency from a single decision point for an entire campaign, then your only exposure to over-frequency during a campaign will be cookie deletion, so you might get a 20 % excess over the course of a 6 week campaign. Estimates vary. If you are running on Facebook, and only Facebook, you will get perfect frequency control. If you are running a nominal multi publisher multi network campaign with only publisher local frequency control, *the number will be more like 40 % but could be as high as 70 %*.

Keep in mind that we universally ask the wrong question regarding frequency. What we usually ask is: What was the average frequency? We should be asking what portion of the viewers received the correct frequency. It is possible, and normal, that most users get one, while others get 30. If the spec was four, it might be that almost no one got four, even if the average was four.

In one Panel based study looking at which impressions were above the specification for a typical direct purchased campaign with 12 premium publishers for a large advertiser, the number was close to 40 %. In other words, 40 % of the impressions were wasted because they were over the frequency specified in the IO/booking.

The nominal assumption in the model is that 60 % of impressions were in inside the frequency specification. This is reasonable based on several kinds of research and several types of companies, per above ... and very generous if we are counting the number of portion of users that got the right number of impressions. Certainly, everyone can agree that 40 % of the impressions in most campaigns are over frequency in today's marketplace. The model allows the user to put in any number they like, however.

4.6 Non-Human Traffic

It's important to realize here that when a server running a browser simulator, rather than a browser with a human in front of it "views" an ad, it's wasted, *regardless* of whether the intention was fraud. There are thousands, perhaps millions of Web Robots. Most have innocent purposes, and many are extremely important to managing the web. This can be detected much of the time, but is difficult to prevent.

Another aspect if this type of waste is that it is site specific. Some sites have less robotic traffic, and some have more.

A third aspect is that detection is an arms race. When we get better at detecting robots, the less benign robots find ways of getting smarter too. This happens constantly. For example, when bot detection learned that bots move cursors in a straight line, bots adapted, using human-like cursor movement. It is unlikely that any "final" solution will be created, so one should view Robotic Traffic as another endemic source of waste, varying by site and browser population. Also, benign bots are helpful to the ecosystem, gathering data about content, engagement, and so on to use to improve advertising results.

How much? Estimates of bot traffic vary between 10 % and 90 % site to site. A better range would be 20–100 (100 % for bot created content), but the average for a well-managed campaign with a white listed publisher might be closer to 25 %. Keep in mind once again that fraudulent impressions are a smaller group than bot impressions because many or most impressions that go to bots are not fraudulent. A recent ANA study identified that 17 % of exchange impressions were fraudulent, so that would define an absolute minimum average across a wide range of exchange-traded inventory.

The nominal assumption in the model is that 25 % of impressions go to robots, thus 75 % "human". Obviously, there are no bots on TV, and very few on sites that require a password to access an account.

4.7 On Target

In mass media, for products like toothpaste, it has been shown for TV that off-target media sometimes provides more lift than the on-target portion of the campaign. It's apparent why. If the Brand is toothpaste, everyone has teeth, but the target is a subset. Simply, the targeting is not so critical, and the commercial reaches way more people than are inside the target. TV buyers buy mass reach this way sometimes because over-delivery is the outcome … and the off target (free) impressions resulting from over delivery do a lot of good when the target specification is narrower than the set of potential buyers.

This is usually not the way it works in the Internet.

Outside of sites with accurate registration data, targeting relies on inference, based on behaviors, as reflected by cookies, or other forms of ID such as Mobile fingerprinting. For example, if a site skews heavily male, it is inferred that visitors to that site are male. In several studies by both Nielsen and Comscore, this results in gender skew at 65–70 % ... when a wild guess will get you 50 %. Of course, when the target incidence is lower, the cookie accuracy is lower (inference less true) too.

When no inference is required to peg the meaning of a cookie, results can be stellar. For example, retargeting shopping carts that abandoned a particular site will yield perfect targeting because a cookie was set to document a specific event for a specific browser. The issue here is that reach may be low, but then so are media costs. In general retargeting 1st- and 2nd-party data is accurate assuming not too much liberty is taken extrapolating the behavior into meaning.

To assume someone intends to buy a car simply because they visited a car site is an example of a stretch.

So called "lookalike models" also expand reach at the expense of target accuracy. In these models, retargeted segments are modeled to create larger segments of cookies that have behaviors similar to the cookie that demonstrated the desired behavior ... such as bought a car. As the model extends reach, the likelihood that the cookie is actually the right target will be lower. For example, the model might notice that people who buy a certain kind of car are slightly richer, slightly older, prone to surf the web early in the morning, and visit insurance sites a lot. Then, from larger data pools, cookies that exhibit these characteristics can be identified. However, the characteristics are an inference to begin with; there are a lot of things that richer older people might like to buy that are not a car, and not all old rich people are in the market for a car.

In the model, for TV and registration sites, we use default high values for reach to target. TV because of how it is delivered (they buy reach until all the targets have been hit statistically), and registered sites for reasons we talked about before. For other cookie or context targeted (normal web publisher or exchange) media, we use 60–70 % (or 30 % wasted) ... the user is free to change these numbers in any way that reflects their anticipated campaign performance.

Something to keep in mind here is that retargeting does not eliminate fraud (we get perfectly targeted bots!) because bots now intentionally visit sites so that they will be "cookied" to appear to be a high value consumer. This makes the price of the fraudulent impressions go up, ergo revenue to the fraudsters.

5 Model Specifics and Calculations

5.1 Model Will Account for the Following Types of Media but Others Can Be Added

- Over the air TV
- Ad networks/eco-system Video
- Network Guaranteed Display
- Normal Exchange traded Display and Video

5.2 Pricing Assumptions in the Model

We have preloaded the model with pricing. The pricing is nominal. By "nominal" we mean we guessed a reasonable average from interviewing advertisers, but there are wild swings (±50 % or more) depending on contract terms, and value chain factors. Users should input their own pricing.

5.3 Treatment of Interactions of Impairments

We call any factor that causes waste inside a population of impressions *impairment*. For example, non-human traffic is impairment to productive media.

The portion of impaired impressions is the percent wasted. The inverse of that is the percent productive. Those impairment percentages add up to more than 100 because many impressions suffer from multiple impairments. For example, a non-human impression might be over frequency, *and* not in view. We deal with these interactions in the model in the following way.

If a population of impressions suffers from two impairments, the chance an impression will suffer from both of them is the product of the probabilities. In effect, we subtract out the chance for both so we don't double count. This method assumes normal distributions of impairments across campaigns, which is likely not the case, but we don't have measures of composite effects of impairments, so it's a pragmatic assumption. Because of the high number of heuristics in the model, small changes in any particular one don't change the outcome much.

6 An Example of the Model in Use

Media Examples (nominal)	# imps	Nominal (Buy) CPM	Total cost in $	In-view	Human	In Geo	On Target	Inside right frequency	ad loaded	Portion of views 70% complete	Net Probability across all impressions	actual quality exposures	Quality E-cpm	Index vs Buy CPM
Over the Air Linear TV	12,000,000	$20.00	$240,000	100%	100%	100%	65%	80%	100%	90%	46.8%	5,616,000	$42.74	214
Guarenteed Display	12,000,000	$10.00	$120,000	90%	90%	80%	70%	70%	95%	100%	30.2%	3,619,728	$33.15	332
Nominal Ex traded Display	12,000,000	$2.00	$24,000	60%	75%	80%	60%	80%	95%	100%	16.4%	1,969,920	$12.18	609
Nominal Ex traded Video	12,000,000	$12.00	$144,000	60%	75%	80%	60%	80%	95%	70%	11.5%	1,378,944	$104.43	870
Nominal Private Exchange	12,000,000	$6.00	$72,000	70%	80%	80%	70%	70%	95%	100%	20.9%	2,502,528	$28.77	480

The static instantiation of the model above compares TV, Online Display in some typical products, and Online Video as available in several typical products. Typical exchange traded display comes out roughly 30 % less expensive than private exchange display because the 3x price difference (assumed) is not compensated by the aggregated effects of differences in assumed impairments.

On the other hand, exchange traded video, which has both a high CPM, and typical impairment profile looks very expensive (almost 3000 %!) compared to

Facebook (FB) video in which impairments are almost non-existent ... and that assumes some non-human traffic on Facebook. In the provided model, all these assumptions can be changed. However the implication remains clear: Negotiating off-rate-card increments (i.e. "10 % off") based on "scale" is almost meaningless compared to the effect of attention quality.

7   Conclusion

Buyers are much better off to architect their systems for quality than to build on organizational principles that commoditize media in hopes that competitive bidding will give them a lower price. Leveraging programmatic technology allows advertisers and agencies to do both – continue to leverage their buying clout through scale and/or other levers *and* harness the optimization opportunities described above.

Mandates and Implications
1. Instead of pixels and time, advertisers should manage toward attention quality and outcomes by implementing any and all measures that provide insight.
2. Advertisers should strongly consider managing frequency by creating a single decision point for their campaigns. Watch Out is the today substantial cost of programmatic tools, which cost up to 30 % of total media value across publishers, agencies and advertisers.
3. Advertisers should be certain that attribution models do not incorporate meaningless impressions. They are noise, not signal.
4. Advertisers should put their actual data into the provided model, and use the outcome in their planning process.
5. Advertisers should not lose sight of factors beyond quality of audience/attention and reach (i.e. contextual fit etc.).

Ted McConnell is a Consultant and Digital Marketing Generalist: he has run large organizations, been staff to c-level executives at Procter & Gamble, hold four patents and successfully drive change in Marketing and Advertising practice in P&G, where he was Manager Digital Marketing Innovation for almost a decade. He is serving or has been recently serving on advisory boards for several successful Advertising Technology companies.

 **Lothar Hoecker** has previously held local and global roles at Procter & Gamble, where he was in a leading role to invent and set the path for P&Gs programmatic advertising future around the globe. Since 2014 Hoecker takes advantages from his vast international experiences in programmatic and non-programmatic media as Germany's Head of Investment Management at DENTSU AEGIS NETWORK. In this role he leads the Strategic Buying across all media and investment.

For Social Good

At the very end a special thank you to all the authors, first for their excellent contribution, and second for joining me in waiving our author's fees in favor of the remarkable charity and education project

CHILDREN'S SHELTER FOUNDATION

www.facebook.com/childrensshelterfoundation

The Children's Shelter Foundation in Chinag Mai is a farm that gives minor orphans, deaf-mute and other disadvantaged children from Northern Thailand & Burma an education, a family and a future.

Printed by Printforce, the Netherlands